P203 更正为

Indarctos sp.
印度熊（未定种）
新近纪，甘肃

Neogene; Gansu
NIGPAS Collection #MV172

P208 更正为

Elephas namadicus (3rd molar)
纳玛象（第三臼齿）
第四纪，江苏南京

Quaternary; Nanjing, Jiangsu

P209 更正为

Serridentinus sp.
锯齿象（未定种）
中中新世，甘肃

Middle Miocene; Gansu

Evolutionary Treasures

岁月菁华

Fossil Types Annotated

化石档案与故事

中国科学院南京地质古生物研究所 编
Edited by Nanjing Institute of Geology and Palaeontology,
Chinese Academy of Sciences

杨 群 主编
Chief Editor YANG Qun

《岁月菁华》编写组

主　编

杨　群

成　员

顾元达　张允白　陈孝政　傅　强　樊晓羿　蔡华伟　毛礼米　冯伟民　袁道俊
盛　捷

摄　影

樊晓羿　傅　强　汤晶晶

供　图 (按姓氏汉语拼音排序)

蔡晨阳　曹美珍　陈　雷　陈　哲　陈均远　李保华　林彩华　刘　锋　罗　辉
祁玉平　孙卫国　谭　超　唐　卿　唐玉刚　王　博　王　军　武　雯　徐　宾
徐洪河　许　波　杨定华　殷宗军　尹磊明　袁训来　曾　晗　张小萍　张元动
赵方臣　钟石兰　周志炎　朱茂炎

前 言

中国科学院南京地质古生物研究所（简称南京古生物所）是专门研究古生物的国家级科研机构，成立于1951年，其前身是前中央研究院地质研究所及前中央地质调查所等机构的古生物室(组)。目前，它是我国唯一从事古生物学（古无脊椎动物学与古植物学）和地层学研究的专业机构，被誉为"国际三大古生物研究中心之一"。

古生物见证了生命起源和生物演化历史。古生物化石以及产出化石的岩石地层是南京古生物所科学家的主要研究对象。地层就好像是一本书，记录了地球的历史，化石则保存了地质历史上生命演化的历程，见证了地球表面沧海桑田、气候变化、火山爆发、板块运动、天地撞击等地质历史事件，也记录了地球表面生态系统的变迁，物种起源、灭绝、更替及其与自然环境协同演化的历史。

南京古生物所几代科学家为了中国的古生物学科发展，历经艰辛，在门类古生物学、地球早期生命的起源与寒武纪大爆发、澄江生物群、热河生物群、重大地史时期生物起源 – 辐射 – 灭绝与复苏、生物地层学与全球标准层型剖面和点位（俗称"金钉子"）、青藏高原等中国各区域和南极综合科学考察等领域取得了丰硕成果。他们在 Science、Nature、PNAS 等国际著名学术刊物上发表的研究成果，多次被评为"中国基础研究十大新闻"、"中国十大科技进展新闻"、"中国科学十大进展"等；他们主持完成的"澄江生物群与寒武纪大爆发"荣获 2003 年国家自然科学奖一等奖。在中国现已建立的 10 个"金钉子"中，有 7 个是由南京古生物所科学家主持完成的。

南京古生物所英才辈出，建所以来先后有李四光、斯行健、赵金科、王钰、卢衍豪、穆恩之、李星学、顾知微、盛金章、周志炎、戎嘉余、金玉玕、陈旭和沈树忠等 14 位科学家当选为中国科学院院士（学部委员）。研究所的科学家先后承担和领导了多项重大国际合作项目，在国际古生物协会（IPA）和国际地层委员会（ICS）及其分会等诸多国际学术组织中担任主席、副主席、选举委员，并在中国多次主办具有重要影响的国际学术会议。

南京古生物所收藏了自中国古生物学诞生之日起，古生物学家近百年来在国内外采集的古生物化石。其中，已经研究、描述、图示并公开发表各类模式标本 16 万余件，收藏着著名学者如葛利普、李四光、孙云铸、尹赞勋等在 1949 年之前采集研究的标本，汇集了研究所建所以来在承担中国各门类化石、中国各纪地层划分对比和各纪地层界线层型剖面研究、青藏高原及南极地层古生物、澄江生物群、热河生物群等重要科研项目中取得的模式标本。

Preface

Fossils are direct evidence of evolution. Palaeontologists explore the fossil world in order to decode the history of biological evolution on the Earth, to understand the interaction of the biosphere and the physical environment in deep time, and to search for the roots of modern biosphere and indicators of the impact of human beings as a species to the ecosystem surrounding our society. In this atlas, we will show you beautiful assorted fossil specimens, which can silently tell the stories that happened on the Earth millions or even billions of years ago, all archived in Nanjing Institute of Geology and Palaeontology, Chinese Academy of Sciences (NIGPAS), mostly collected by NIGPAS scientists from different regions of China and elsewhere.

NIGPAS is widely known as one of the three major palaeontological research institutions in the world. The Institute was founded in 1951 with its roots extending to the former palaeontological laboratories in the National Research Institute of Geology (Academia Sinica) and the National Geological Survey of China before 1949.

NIGPAS excels in studies of invertebrate animal fossils, fossil plants and pollen-spores, microfossils (very small and better observed under microscopes), biostratigraphy (studying rock sequences using fossil occurrences and associated data), chronostratigraphy (time division and correlation of rocks using various criteria such as biozones, radiometric dates, geochemical signatures, etc.). NIGPAS houses modern research facilities including laboratories of palaeobiology, organic and inorganic geochemistry, sedimentology and data analysis, a library and information center, a type fossil repository, and a field station. We also have several public outreach platforms, including the Nanjing Museum of Palaeontology, a popular magazine *Evolution of Life* (《生物进化》) and a social cyberspace "Fossil Web" (化石网) with over 135,000 members.

NIGPAS has treasured generations of active palaeontological researchers who have devoted their life time in searching for fossils in all China to as far as the Antarctica and in working hard in laboratories with tremendous efforts to try to decode the meanings of the fossils — biological relicts from the earth history before the human being came along. Fourteen of the faculty members have been elected as academicians (or Members of the Earth Science Division) in Chinese Academy of Sciences, all being leading experts in relevant research areas worldwide, among many other well-known palaeontologists and stratigraphers, who contributed to the advancement of palaeontological discipline in China and to the world.

2005 年，南京古生物博物馆建成并向社会正式开放。博物馆展览以精美的古生物化石和科研成果为主体，包括"澄江生物群"、"热河生物群"等特异埋藏化石群中的珍贵标本，如奇虾、微网虫、中华龙鸟、雌雄孔子鸟、辽宁古果等化石标本，堪称国宝级化石精品。

岁月荏苒，菁华凝练。本图集向大家展示的，既有南京古生物所收藏的精品化石，也有南京古生物所科学家在探索生命演化历史的历程中的成果实例。每件化石的背后都有着精彩的故事；它们不仅再现了几十亿年来地球生命的恢弘演化史，还记录了中国古生物学充满艰辛曲折却又辉煌灿烂的百年发展史。

感谢科技部、国家自然科学基金委、国土资源部、中国科协、中国科学院、江苏省和南京市人民政府以及社会各界对南京古生物所的关心和支持！感谢国内外各兄弟单位和个人的支持与合作。

本图集编撰工作得到南京古生物所各位专家学者和各部门的大力支持；中科院古脊椎动物与古人类研究所邓涛研究员、汪筱林研究员在脊椎动物化石方面给予指导；部分历史素材源于《中国古生物学学科史》（中国科学技术出版社，2015）和南京古生物所未发表资料。由于编写人员的水平及编写时间所限，不足之处敬请批评指正。

杨　群　博士

中国科学院南京地质古生物研究所 所长
中国古生物学会 理事长
2017 年 7 月

This atlas displays a good sample of beautiful fossil collections in NIGPAS, including most precious specimens collected by pioneering palaeontologists in early 1900s and exquisitely preserved fossil specimens from the famous "Chengjiang Biota" and "Jehol Biota", as well as other important fossil sites of various geological ages and geographic regions. This atlas is intended not only to show snap shots of the evolving biosphere on the Earth through millions and billions of years' natural history, but also to show some footprints of NIGPAS scientists in their over 60 years' scientific endeavors (plus pre-NIGPAS pioneers' efforts). Their research achievements have widely received international and national recognition, many published in major academic journals such as Science, Nature and PNAS and granted the top awards for basic research in China.

Major funding of our research has come from Ministry of Science and Technology, National Natural Science Foundation, Chinese Academy of Sciences, China Association of Science and Technology, Ministry of Land and Resources (and the former Ministry of Geology and Mineral Resources), and the local governments of Jiangsu Province and Nanjing City. The collaborations with numerous researchers and field explorers throughout China and from other countries have been vital and necessary for all the achievements at NIGPAS.

The compiling team of this atlas would like to thank the support and contributions, by consultation, assistance in photography or material supply, from their colleagues of departments and offices at NIGPAS and from Prof. Dr. Deng Tao and Prof. Dr. Wang Xiaolin of the Institute of Vertebrate Paleontology and Paleoanthropology (CAS) in Beijng. Some of the historical accounts are based on the book *Chinese History of Palaeontology* (Press of Science and Technology of China, 2015) and unpublished documents from NIGPAS.

Prof. Dr. Yang Qun
Director of NIGPAS
President, Palaeontological Society of China
July, 2017

目 录
CONTENTS

史海掠影
Historical Spot Lights / 001

前寒武纪
Precambrian Life Forms / 037

古生代
Palaeozoic Animals and Plants / 053

中生代
Mesozoic:
The Lost World and the Forerunners of the Modern Ecosphere / 135

新生代
Cenozoic:
Era of New Life after Ammonites and Dinosaurs / 195

后 记
Postscript / 220

史海掠影

Historical Spot Lights

Evolutionary Treasures

古代中国的化石记录 / Fossils Known to Ancient Chinese

沈括 Shen Kuo
(1031–1095)

据文献记载，中国最早关于化石的记录出现在两千多年前春秋战国时期的《山海经》中，其中提到的"龙骨""龙鱼"等就是脊椎动物化石。

北宋学者沈括在《梦溪笔谈》中首次明确表述化石的来源和本质（他所描述的"竹笋"经考证很可能是新芦木类化石），认为化石是经历沧海桑田，最终存留在岩石中的古生物遗体、遗物或生活痕迹。

In ancient China, fossils were noted in various literatures tracing back to the Warring States (475 B.C.–221 B.C.) to the Han Dynasty (206 B.C.–220 A.D.) in the monumental works "Shan Hai Jing", in which "dragon bones" and "dragon fishes" were included as part of the natural resources in the volumes of encyclopedic texts and atlas. During the Northern Song Dynasty (960–1127), the famous scholar Shen Kuo (1031–1095), in his encyclopedic magnum opus "Meng Xi Bi Tan", described the fossil plant from northern Shaanxi Province as "stone bamboos", exposed due to the collapse of a river bank. These "stone bamboos", according to their occurrence, are now assigned to *Neocalamites*, extinct tree-like horsetails of Triassic and Jurassic times (about 252–145 Ma). Because there were no "bamboo-type" plants in northern Shaanxi, Shen Kuo interpreted this occurrence as evidence of vast environmental changes since the formation of the fossils.

Neocalamites carrerei
卡勒莱新芦木
晚三叠世，陕西宜君

Late Triassic; Yijun, Shaanxi
NIGPAS Collection #PB2256

Calamites suckowii
钝肋芦木
二叠纪，内蒙古清水河

Permian; Qingshuihe, Inner Mongolia
NIGPAS Collection #PB3908

黄庭坚 Huang Tingjian
(1045–1105)

北宋著名诗人、书法家黄庭坚收藏了中国第一件经人工打磨的鹦鹉螺类化石标本——中华震旦角石（*Sinoceras chinense*），并在其上镌刻了他的亲笔题诗："南崖新妇石，霹雳压笋出。勺水润其根，成竹知何日。"

The ancient poet and calligrapher Huang Tingjian (1045–1105) wrote a poem for the fossil nautiloid (now we know it belongs to the widespread fossil species in South China: *Sinoceras chinense*, of Ordovician time more than 450 Ma (million years ago) with his beautiful calligraphy on the well-prepared specimen. This suggests that the ancient Chinese collected and prepared fossils as artworks and this piece may be the earliest fossil artwork well preserved by human beings.

中国古生物研究的先驱 / Pioneers of Palaeontology in China

德国学者李希霍芬 (Ferdinand von Richthofen，1833–1905)

在中国旅行五年，他编著的 5 卷本著作《中国：亲身旅历及研究成果》，包括了古生物学的专论以及志留纪至第四纪地层的综合论述，成为我国地层古生物学发展早期的重要参考书。

Ferdinand von Richthofen spent five years extensively traveling in China, resulted in the publication of 5-volume book series — *China: Ergebnisse Eigener Reisen und Derauf Gegründeter Studien*, which included palaeontological works and stratigraphic syntheses of Silurian through Quaternary periods that later became widely referenced during the early years of the development of Chinese palaeontology and stratigraphy.

李希霍芬 F. von Richthofen (1833–1905)

美国学者葛利普 (Amadeus William Grabau, 1870–1946)

他是对中国地质古生物学贡献最大的外国学者，曾在中国地质调查所和北京大学任教，培养了中国最早一批地层古生物学者。

American scholar Amadeus William Grabau is regarded as the most influential foreign figure to the development of Chinese geology and palaeontology. He taught in Peking University and in the National Geological Survey of China where the first generation of Chinese geologists and palaeontologists were trained.

葛利普 A.W. Grabau (1870–1946)

法国学者德日进 (Pierre Teilhard de Chardin, 1881–1955)

他在 1923—1946 年长期调查中国的地层、古生物和区域地质，并为之作出重要贡献。

French scholar Pierre Teilhard de Chardin contributed to Chinese palaeontology and regional geology through almost 20 years exploration in wide regions of China during 1923–1946.

瑞典学者赫勒 (Thore Gustaf Halle, 1884–1964)

他于 20 世纪早期应邀在中国地质调查所工作，对中国古植物学的建立和发展贡献良多。

Swedish scholar Thore Gustaf Halle was instrumental in the development of Chinese palaeobotany as an advisor to the National Geological Survey of China in the 1910s.

Evolutionary Treasures

李希霍芬的《中国》及植物化石图版

Five volumes of the works *China: Ergebnisse Eigener Reisen und Derauf Gegründeter Studien*, with one of the plates of fossil plants, by Ferdinand von Richthofen

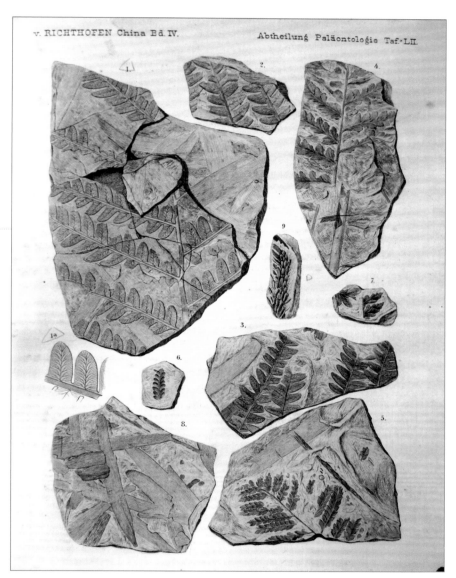

Stereoplasmoceras pseudoseptatum
假隔壁灰角石
中奥陶世，河北唐山滦县
葛利普1922年研究发表

Middle Ordovician; Luanxian, Tangshan, Hebei
Published by A.W. Grabau (1922)
NIGPAS Collection #58

Evolutionary Treasures

章鸿钊（1877—1951），浙江湖州人，1905年留学日本，入东京帝国大学地质学系学习；1911年毕业回国，在京师大学堂农科任地质学讲师。中国地质事业创始人和奠基人之一，中国地质学会首任会长。

Zhang Hongzhao (H.T. Chang, 1877–1951), born in Huzhou of Zhejiang, graduated from Department of Geology, Tokyo Imperial University, Japan in 1911. He taught geology in the Imperial University of Peking and became one of the key founders of Chinese geology. He was the founding president of Geological Society of China.

丁文江（1887—1936），江苏泰兴人，1902年留学日本，1904年留学英国，1911年毕业于格拉斯哥大学，获地质学、动物学双学士学位。中国地质事业创始人和奠基人之一，牵头创办中国第一个地质机构——中国地质调查所。

Ding Wenjiang (V.K. Ting, 1887–1936), born in Taixing of Jiangsu, majored in zoology and geology in University of Glasgow (Britain) during 1907–1911. He was also one of the key founding scientists of Chinese geology. In 1916, he became the founding director of the National Geological Survey of China.

丁文江 Ding Wenjiang
（1887–1936）

翁文灏（1889—1971），浙江鄞县人，1908年留学比利时鲁汶大学地质系，1912年毕业，在23岁时成为中国地质学界第一位博士。中国地质事业创始人和奠基人之一，中国第一张着色全国地质图的编制者。

Weng Wenhao (W.H. Wong, 1889–1971), born in Yinxian of Zhejiang, studied in Department of Geology, University of Leuven (Belgium) during 1908–1912 and received the first doctoral degree in geology among Chinese scholars. He was also one of the key founders of Chinese geology; he compiled the first colored geological map of China.

Pecopteris sp.
栉羊齿（未定种）
二叠纪，云南宣威
丁文江采集，T.G. Halle 1927年研究发表

Permian; Xuanwei, Yunnan
Collected by Ding Wenjiang
Published by T.G. Halle (1927)
NIGPAS Collection #PB13

Protolepidodendron scharyanum
夏利安原始鳞木
早泥盆世，云南曲靖
丁文江采集，T.G. Halle 1936年研究发表

Early Devonian; Qujing, Yunnan
Collected by Ding Wenjiang
Published by T.G. Halle (1936)
NIGPAS Collection #PB119

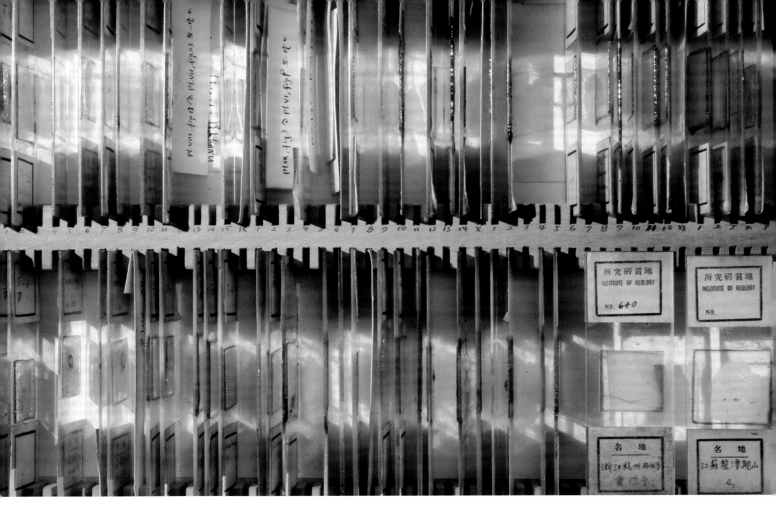

李四光（1930）研究的蜓类薄片

Thin sections of fusulinids
filed by Li Siguang in 1930
NIGPAS Collection #14–905

中国地质古生物学家中最早发表古生物学论文的是李四光（1923）、周赞衡（1923），最早发表古生物学专著的是孙云铸（1924）。中国古生物学系统研究中的主要化石门类及早期主要学术带头人有：蜓类（李四光、陈旭、盛金章）、珊瑚（乐森璕、黄汲清、俞建章、计荣森、王鸿祯）、苔藓动物（乐森璕）、腕足动物（赵亚曾、黄汲清、田奇㻪）、头足类（俞建章、尹赞勋、田奇㻪、孙云铸、许德佑）、腹足类（秉志、尹赞勋、许杰、阎敦建）、双壳类（赵亚曾、许德佑、顾知微）、三叶虫（孙云铸、卢衍豪、张文堂）、棘皮动物（田奇㻪、孙云铸、穆恩之）、笔石（孙云铸、许杰、穆恩之）、古脊椎动物与古人类（杨钟健、裴文中、贾兰坡、卞美年、刘东生、周明镇、吴汝康、刘宪亭）、古植物（周赞衡、斯行健、李星学）（引自《中国古生物学学科史》，中国科学技术出版社，2015）。

Among the pioneering Chinese geologists and palaeontologists, Li Siguang (1923) and Zhou Zanheng (T.C. Chow) (1923) were the first to publish palaeontological research papers (on fusulinids and fossil plants, respectively); Sun Yunzhu (Y.C. Sun) (1924) was the first to publish palaeontological monograph (on Cambrian fossils). Pioneering or major contributing fossil specialists in the history are: Li Siguang, Chen Xu (X. Chen), and Sheng Jinzhang (J.C. Sheng) on fusulinids; Yue Senxun (S.S. Yoh), Huang Jiqing (T.K. Huang), Yu Jianzhang (C.C. Yu), Ji Rongsen (Y.S. Chi), and Wang Hongzhen (H.C. Wang) on corals; Yue Senxun on bryozoans; Zhao Yazeng (Y.T. Chao), Huang Jiqing, and Tian Qijun (C.C. Tien) on brachiopods; Yu Jianzhang, Yin Zanxun (T.H. Yin), Tian Qijun, Sun Yunzhu, and Xu Deyou (T.Y. Hsu) on cephalopods; Bing Zhi (C. Ping), Yin Zanxun, Xu Jie (C. Hsu), and Yan Dunjian (T.C. Yen) on gastropods; Zhao Yazeng, Xu Deyou, and Gu Zhiwei (C.W. Ku) on bivalves; Sun Yunzhu, Lu Yanhao (Y.H. Lu), and Zhang Wentang (W.T. Chang) on trilobites; Tian Qijun, Sun Yunzhu, and Mu Enzhi (A.T. Mu) on echinoderms; Sun Yunzhu, Xu Jie, and Mu Enzhi on graptolites; Yang Zhongjian (C.C. Young), Pei Wenzhong (W.C. Pei), Jia Lanpo (L.P. Chia), Bian Meinian (M.N. Bien), Liu Dongsheng (T.S. Liu), Zhou Mingzhen (M.Z. Chou), Wu Rukang (R.K. Wu), and Liu Xianting (H.T. Liu) on vertebrate palaeontology and palaeoanthropology; Zhou Zanheng, Si Xingjian (H.C. Sze), and Li Xingxue (H.H. Lee) on fossil plants (*Chinese History of Palaeontology*, Press of Science and Technology of China, 2015).

Evolutionary Treasures

Pagiophyllum sp.
坚叶杉（未定种）
白垩纪，山东莱阳
此为中国学者研究的第一批古植物标本，由周赞衡发表于1923年

Cretaceous; Laiyang, Shandong
Studied by Zhou Zanheng (1923), considered to be the first fossil plant specimen ever studied by a Chinese researcher
NIGPAS Collection #PB153

Pseudosageceras? sp.
假胄菊石？（未定种）
三叠纪，湖北保安
标本研究者许德佑1944年在贵州野外考察期间被土匪杀害

Triassic; Bao'an, Hubei
Studied by Xu Deyou, who was killed along with two other geologists by robbers during a field survey in 1944
NIGPAS Collection #3495

Dictyclostus taiyuanfuensis
太原网格长身贝
晚石炭世，山西太原关底沟

Late Carboniferous; Guandigou, Taiyuan, Shanxi
NIGPAS Collection #1020

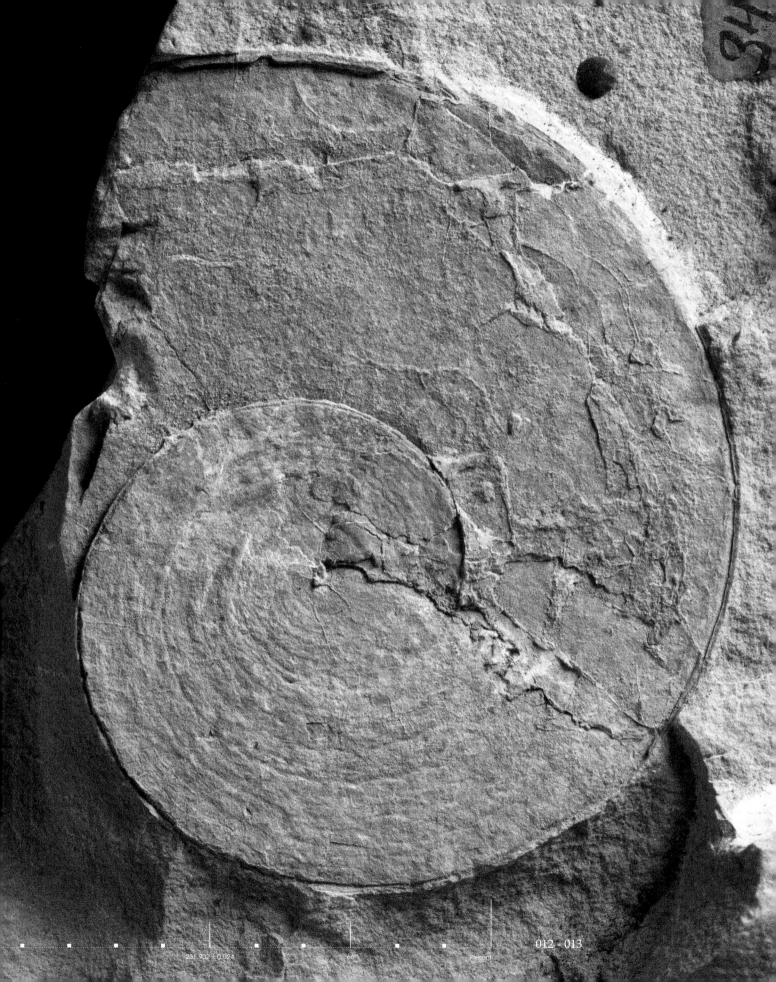

012 – 013

农商部地质调查所图书馆陈列馆开幕典礼摄影（1922年）

第一排左三谢家荣，左四王竹泉，左五袁复礼；第二排左一葛利普，左二谭锡畴，右二章鸿钊，右三丁文江；第三排正中是黎元洪，右墙站立者邢端，旁边是周赞衡。

Opening ceremony of the Library of Geological Survey, Ministry of Agriculture and Commerce (1922)

Geologists seated in front row: Xie Jiarong (3rd left), Wang Zhuquan (4th left), Yuan Fuli (5th left); seated in 2nd row: A.W. Grabau (1st left), Tan Xichou (2nd left), Zhang Hongzhao (2nd right), Ding Wenjiang (3rd right); seated in 3rd row middle: Li Yuanhong (President of the Republic of China).

Evolutionary Treasures

国立北京大学地质学会欢送一九三六班毕业纪念

Graduation ceremony for geology students at National Peking University (1936) joined by members of the Geological Society

中国古生物学会复活大会会员摄影 (1947年)

Resurrection Meeting (1947) of the Palaeontological Society of China after WWII

Evolutionary Treasures

中国科学院古生物研究所的诞生及变迁

1950 年，中国科学院决定在前中央研究院地质研究所及前中央地质调查所等机构的古生物室（组）的基础上，筹建中国科学院古生物研究所。1950 年 8 月 25 日，政务院总理周恩来签发任命书，李四光任首任所长。1951 年 5 月 7 日，中国科学院古生物研究所正式成立，内设古植物组、古无脊椎动物组和古脊椎动物组。

第一批研究人员包括斯行健、杨钟健、赵金科、俞建章、卢衍豪、王钰、徐仁、穆恩之、李星学、顾知微、裴文中、杨敬之、周明镇、贾兰坡、刘宪亭、刘东生、盛金章、侯祐堂、张文堂、王水、胡长康（引自《中国古生物学学科史》，中国科学技术出版社，2015）。

1953 年 4 月，在北京的古脊椎动物研究室由中国科学院直接领导，后成立中国科学院古脊椎动物研究所。

1959 年 5 月更名为中国科学院地质古生物研究所；1971 年 3 月更为现名——中国科学院南京地质古生物研究所。目前，研究所是中国唯一从事古生物学（古无脊椎动物学与古植物学）和地层学研究的国家级专业科研机构，被誉为"国际三大古生物研究中心之一"。

中国科学院古生物研究所建所大会（1951年）
Founding congress of CAS Institute of Palaeontology (1951)

Founding of the Institute of Palaeontology, Chinese Academy of Sciences

In May 1951, the Institute of Palaeontology, Chinese Academy of Sciences (CAS), was formally established in Nanjing, with its laboratories and staff members derived from the former palaeontological laboratories of Academia Sinica and the National Geological Survey of China (GSC), together with the Cenozoic Research Department of the GSC Peking Branch.

Li Siguang was appointed the first director of the institute by Premier Zhou Enlai, with Si Xingjian as the acting director and Zhao Jinke and Lu Yanhao as the vice directors. The founding faculty members also included: Yang Zhongjian, Yu Jianzhang, Wang Yu, Xu Ren, Mu Enzhi, Li Xingxue, Gu Zhiwei, Pei Wenzhong, Yang Jingzhi, Zhou Mingzhen, Jia Lanpo, Liu Xianting, Liu Dongsheng, Sheng Jinzhang, Hou Youtang, Zhang Wentang, Wang Shui, and Hu Changkang (*Chinese History of Palaeontology*, Press of Science and Technology of China, 2015). In April 1953, the Department of Vertebrate Palaeontology (stationed in Beijing) became directly affiliated with CAS in administration and consequently the CAS Institute of Vertebrate Paleontology and Paleoanthropology (IVPP) was established in Beijing.

The CAS Institute of Palaeontology was renamed CAS Institute of Geology and Palaeontology in May 1959, and again changed its name in March 1971 to become Nanjing Institute of Geology and Palaeontology, CAS (NIGPAS), which is the current institute name. It is currently the only national institution specialized in invertebrate palaeontology and palaeobotany (including micropalaeontology and palynology), known to be one of the three major palaeontological research centers in the world.

Evolutionary Treasures

周恩来总理（右）同李四光教授亲切交谈（1952年）
Li Siguang greeted by Premier Zhou Enlai (1952)

李四光 Li Siguang
（1889–1971）

李四光（1889—1971），湖北黄岗人，1904年被官派留学日本，1910年毕业于大阪高等工业学校，1931年获英国伯明翰大学博士学位，1955年当选为中国科学院学部委员（院士）。他是中国地质事业的奠基人之一和主要领导者，牵头创办中央研究院地质研究所和中国科学院古生物研究所，首创汉字"蜓"及蜓科化石分类鉴定标准。

Li Siguang (J.S. Lee) (1889–1971), born in Huanggang of Hubei, sent to Japan for higher education by government, graduated from Osaka Industrial College (Japan) in 1910, and obtained doctoral degree at University of Birmingham (UK) in 1931. He is regarded as one of the founders and leaders of early Chinese geology. He was the founding director of the National Research Institute of Geology (Academic Sinica) and CAS Institute of Palaeontology. He coined the Chinese character for fusulinids and established a new classification system for the fusulinid family. He was elected to CAS in 1955.

斯行健 Si Xingjian
（1901–1964）

斯行健（1901—1964），浙江诸暨人，1926年毕业于北京大学地质系，1931年获德国柏林大学博士学位，1953年任中国科学院南京古生物所所长，1955年当选为中国科学院学部委员（院士）。他在古植物学的众多领域做出了开拓性工作，奠定了中国古植物学和陆相地层研究的基础。

Si Xingjian (H.C. Sze) (1901–1964), born in Zhuji of Zhejiang, graduated from Department of Geology, Peking University in 1926, and received a doctoral degree at Berlin University (Germany) in 1931. He served as NIGPAS director from 1953 to 1964 and was elected to CAS in 1955. He pioneered in areas of Chinese palaeobotany and terrestrial stratigraphy.

Evolutionary Treasures

赵金科 Zhao Jinke
(1906–1987)

赵金科 (1906—1987)，河北曲阳人，1932年北京大学地质系毕业，1937—1939年在美国哥伦比亚大学深造，1965年任中国科学院南京古生物所所长，1980年当选为中国科学院学部委员（院士）。他在构造地质学、矿产地质学和工程地质学等领域卓有建树，为中国古生物学研究作出了重要贡献，是中国头足类学科的奠基人之一。

Zhao Jinke (K.K. Chao) (1906–1987), born in Quyang of Hebei, graduated from Department of Geology, Peking University in 1932 and received further training in palaeontology at Columbia University (USA) during 1937–1939. He served as NIGPAS acting director and then director from 1965 to 1984 and was elected to CAS in 1980. As one of the founders of Chinese cephalopod palaeontology, he also made outstanding contribution to the research of tectonics and exploration of mineral resources.

卢衍豪 Lu Yanhao
(1913–2000)

卢衍豪 (1913—2000)，福建永定人，1937年北京大学地质系毕业，1945—1946年在美国地质调查局考察学习，1980年当选为中国科学院学部委员（院士）。他奠定了中国寒武纪、奥陶纪地层建阶和分带基础，在三叶虫研究方面颇有造诣，创立了"生物—环境控制论"学说，为中国轮藻化石研究奠定了基础。

Lu Yanhao (Y.H. Lu) (1913–2000), born in Yongding of Fujian, graduated from Department of Geology, Peking University in 1937, and studied in US Geological Survey during 1945–1946. He was elected to CAS in 1980. He made fundamental contribution to the establishment of Chinese Cambrian and Ordovician stratigraphy and the division in stages and biozones. He excelled in the studies of trilobites and laid foundation for charophyte research in China. He proposed the Bio-Environmental Control Hypothesis.

穆恩之 Mu Enzhi
(1917–1987)

穆恩之（1917—1987），江苏丰县人，1943 年西南联合大学地质地理气象学系毕业，1980 年当选为中国科学院学部委员（院士）。他是中国笔石学科的带头人，提出了奥陶纪与志留纪的地层划分对比方案，提出编著各门类化石丛书，领导了西南石油会战、西藏综合考察等地层古生物学研究。

Mu Enzhi (A.T. Mu) (1917–1987), born in Fengxian of Jiangsu, graduated from Department of Geology, Geography and Meteorology, Southwest United University in 1943. He was elected to CAS in 1980. He worked as an academic leader in graptolite studies in China, established the subdivisions of Chinese Ordovician and Silurian stratigraphy, and was instrumental in compiling the atlas of major fossil groups in China. He also led the Petroleum Exploration Campaign in Southwest China and the palaeontology and stratigraphy team in the Multidisciplinary Xizang (Tibet) Expedition.

王钰 Wang Yu
（1909–1984）

王钰（1909—1984），河北深泽人，1933 年北京大学地质系毕业，1944—1946 年在美国国家自然科学博物馆进行访问研究，1980 年当选为中国科学院学部委员（院士）。他对扬子区下古生界的研究是中国南方早古生代地层分类与对比的奠基性工作，奠定了中国南方泥盆纪地层研究的基础。他是中国腕足动物学科的奠基人之一。

Wang Yu (1909–1984), born in Shenze of Hebei, graduated from Department of Geology, Peking University in 1933, was a visiting scholar at US National Museum of Natural Science from 1944 to 1946. He was elected to CAS in 1980. As one of the founding scholars of brachiopod studies in China, he made fundamental contribution in Devonian stratigraphy of South China; his extensive studies in the Early Palaeozoic palaeontology and stratigraphy in the lower Yangtze region were fundamental to all the South China regions.

Evolutionary Treasures

李星学 Li Xingxue
(1917–2010)

李星学（1917—2010），湖南郴县人，1942年重庆大学地质系毕业，1980年当选为中国科学院学部委员（院士）。他以研究古植物学及非海相地层学见长，在对华夏植物群，包括大羽羊齿类植物和东亚晚古生代煤系等的研究方面取得突破性成就。

Li Xingxue (H.H. Lee)(1917–2010), born in Chenxian of Hunan, graduated from Department of Geology, Chongqing University in 1942. He was elected to CAS in 1980. He specialized in research on palaeobotany and non-marine stratigraphy. He had breakthrough achievements in the study of Cathaysia flora, including gigantopterids and Late Palaeozoic coal measures of East Asia.

顾知微 Gu Zhiwei
(1918–2011)

顾知微（1918—2011），江苏南京人，1942年西南联合大学地质地理气象学系毕业，1980年当选为中国科学院学部委员（院士）。他率先对中国淡水双壳类化石和中生代非海相地层开展研究，是著名的"热河生物群"研究的奠基人，为指导石油地质勘探和大庆油田的开发作出了贡献。

Gu Zhiwei (C.W. Gu) (1918–2011), born in Nanjing of Jiangsu, graduated from Department of Geology, Geography and Meteorology, Southwest United University in 1942. He was elected to CAS in 1980. He pioneered in freshwater bivalve studies and Mesozoic non-marine stratigraphy in China; he also pioneered in the studies of the famous "Jehol Biota"; he contributed to petroleum explorations and the discovery of Daqing Oilfield.

盛金章 Sheng Jinzhang
(1921–2007)

盛金章（1921—2007），江苏靖江人，1946年重庆大学地质系毕业，1991年当选为中国科学院学部委员（院士）。他主要从事䗴类及二叠纪生物地层学研究，为中国石炭系和二叠系的分统、建阶打下基础，开展了中国上二叠统"长兴阶"的研究，为国际海相二叠系的对比提供了重要依据。

Sheng Jinzhang (J.C. Sheng) (1921–2007), born in Jingjiang of Jiangsu, graduated from Department of Geology, Chongqing University in 1946. He was elected to CAS in 1991. He excelled in fusulinid studies and Permian stratigraphy, made fundamental contribution to the series and stage divisions of Permian stratigraphy in China and their international correlation, leading to the adoption of Changhsingian Stage to the International Chronostratigraphic Chart.

金玉玕 Jin Yugan
(1937–2006)

金玉玕（1937—2006），浙江东阳人，1959年南京大学地质系毕业，2001年当选为中国科学院院士。他主要从事腕足动物化石研究，是中国石炭纪和二叠纪地层研究的学术带头人，首先提出二叠纪大灭绝的两幕式模式和前乐平统海洋动物灾变事件，曾任国际古生物协会副主席。

Jin Yugan (1937–2006), born in Dongyang of Zhejiang, graduated from Department of Geology, Nanjing University in 1959. He served as a vice chairman of International Palaeontological Association and was elected to CAS in 2001. He excelled in Palaeozoic brachiopod studies and Carboniferous–Permian stratigraphy as an academic leader. He proposed a two-stage end-Permian mass extinction event and a pre-Lopingian event.

Evolutionary Treasures

周志炎 Zhou Zhiyan

周志炎，1933年生，上海人，祖籍浙江海宁，1954年南京大学地质系毕业，曾任国际古植物协会副主席，1995年当选为中国科学院院士。他以中生代裸子植物和蕨类化石的研究见长，开拓了古植物的生物学研究新领域，找到了银杏演化的"缺失链环"。

Zhou Zhiyan, born in Shanghai and a native of Haining, Zhejiang Province, graduated from Department of Geology, Nanjing University in 1954. He served as a vice chairman of the International Association of Palaeobotany and was elected to CAS in 1995. He excels in studies of Mesozoic gymnosperms and fern fossils. He has pioneered in new research directions of palaeobotany and found "missing links" in the evolution of Gingkos.

戎嘉余 Rong Jiayu

戎嘉余，1941年生，上海人，祖籍浙江鄞县，1962年北京地质学院古生物专业毕业，1997年当选为中国科学院院士。他主要从事早—中古生代海洋无脊椎动物（腕足动物门）的系统分类、生物地层与古地理等领域的研究，尤其注重生物宏演化研究，曾任科技部"973"项目首席科学家。

Rong Jiayu, born in Shanghai and a native of Yinxian, Zhejiang Province, graduated from Beijing College of Geology in 1962, majoring in palaeontology. He was elected to CAS in 1997. He has made outstanding contribution to the studies of Early Palaeozoic invertebrate fossils (mainly brachiopods), especially on systematics, biostratigraphy and palaeogeography, and also has focused on macroevolutionary studies. He worked as Chief Scientist of a major research project ("973") funded by the Ministry of Science and Technology of China.

陈旭 Chen Xu

陈旭，1936年生，江苏南京人，祖籍浙江湖州，1959年北京地质学院地质调查及找矿系毕业，2003年当选为中国科学院院士。他主要从事中国奥陶纪和志留纪地层学及笔石动物群的古生物学研究，在中国确立了第一个"金钉子"剖面（浙江常山），比较系统地阐述了显生宙气候带的演变。

Chen Xu, born in Nanjing and a native of Huzhou, Zhejiang Province, graduated from Beijing College of Geology in 1959, majoring in geological exploration and mining. He was elected to CAS in 2003. He has made outstanding contribution to Chinese Ordovician and Silurian stratigraphy and graptolite studies, established the first "golden spike" (GSSP, in chronostratigraphy) in China (base of the Darriwillian Stage, Changshan, Zhejiang), and is well known in interpretation of Phanerozoic climatic evolution.

沈树忠 Shen Shuzhong

沈树忠，1961年生，浙江湖州人，中国矿业大学博士，2015年当选为中国科学院院士。他主要从事二叠纪地层学、二叠纪末生物大灭绝与环境变化、腕足动物古生物学等方面的研究，对二叠纪末海陆生物大灭绝的时序和原因等进行了深入阐述，曾任科技部"973"项目首席科学家。

Shen Shuzhong, born in Huzhou of Zhejiang, received his Ph.D. in China University of Mining and Technology. He was elected to CAS in 2015. He has made outstanding contribution to studies of Permian stratigraphy, end-Permian mass extinction event and environment, and brachiopod palaeontology. He was Chief Scientist of a major research project ("973") funded by Ministry of Science and Technology of China.

辉煌成就 / Glorious Achievements

中国科学院古生物研究所在成立之初,承担了中国地质工作计划指导委员会和原地质部组织的全国地质矿产调查。1955 年,研究所工作重心开始转移到古生物科学研究中来,相继出版了《中国标准化石》《中国各门类化石》《全国地层会议学术报告汇编》,以及《中国古生物志》和《古生物学丛书》系列专著等,为指导全国的矿产勘探、普及古生物学及地层学知识发挥了重要作用。

During the early years after the founding of CAS Institute of Palaeontology in Nanjing, scientists devoted to nationwide exploration of geology and mineral resources sponsored by the central government. Their focus started to change to palaeontological researches in 1955, which consequently resulted in the publication (as the main authors) of fundamental palaeontologic and stratigraphic works including

- Book series (5 volumes): *Index Fossils in China* (1954–1957)
- Book series (15 volumes): *Taxonomic Groups of Fossils in China* (1962–1976)
- Book series (19 volumes): *Proceedings of National Stratigraphy Congress Held in Beijing 1959* (1962–1963)

and a series of monographic works published in *Palaeontologia Sinica* and in *Bulletins of Palaeontology*. These works were not only instrumental for the development of palaeontology and stratigraphy in China, but also enthusiastically welcome among field and mining geologists nationwide for their practical use in their explorations.

南京古生物博物馆
Nanjing Museum of Palaeontology

4000   2500   1600   1000   541.0±1.0
Time (Ma)

研究所参与组织召开了第一届全国地层会议（1959），会议首次对我国各纪地层做了系统总结，为建立中国地层规范、地层分区等基础性科研工作奠定了坚实的基础。

为服务国家地质调查和矿产勘探需求，研究所组织并参加了多次大规模科学考察，包括1965—1972年西南地区石油会战，1966—1968年珠穆朗玛峰地区综合科考，1973—1976年西藏地区综合科考，1975—1976年渤海沿岸地区含油气地层及古生物研究、富铁矿会战等。

Scientists at the institute put forward the initiative and organized the First National Stratigraphic Congress in 1959, when the National Commission on Stratigraphy of China was established, setting the foundation for developing Chinese stratigraphic standards, division and nomenclature. NIGPAS scientists also played an important role in government-sponsored large-scale geological expeditions aimed at finding natural resources, including the 1965–1972 Petroleum Exploration Campaign in Southwest China, the 1966–1968 Mt. Qomolangma Expedtion, the 1973–1976 Mutlidisciplinary Expedition to Tibet, the 1975–1976 Bohai Coastal Petroleum-bearing Stratigraphy and Palaeontology and the National Campaign for Exploring Iron-rich Mines.

Evolutionary Treasures

1978 年，中国步入了改革开放的新时期，南京古生物所的发展迎来了新的辉煌。研究所科研人员在早期生命研究与寒武纪大爆发、"澄江生物群"、"瓮安生物群"、"热河生物群"、全球界线层型剖面和点位（俗称"金钉子"）、生物起源－辐射－灭绝与复苏等研究方向上取得系列重要研究成果，先后在 Science、Nature、PNAS 等国际权威学术刊物上发表研究论文数十篇；相关成果多次荣获国家级奖励，并获评"中国基础研究十大新闻""中国十大科技进展新闻""中国科学十大进展"等。研究所科研人员参与完成的"大庆油田发现过程中的地球科学工作"荣获 1982 年度国家自然科学奖一等奖，"青藏高原的隆起及其对自然环境和人类活动影响的综合研究"荣获 1987 年度国家自然科学奖一等奖。

When China reopened to the world in 1978, NIGPAS began to enter a new era of rapid development and international collaborations. Major research advances have been made in numerous research fronts, including the early life evolution and the Cambrian explosion, the sensational discoveries of the "Chengjiang Biota", the "Weng'an Biota", and the "Jehol Biota", the Global Standard Stratotype-sections and Points (GSSPs, also known as the "Golden Spikes"), and the macroevolutionary events (originations, radiations, extinctions and recoveries), which have been widely recognized nationally and internationally with publications in major international academic journals such as Science, Nature and PNAS and received the highest national awards for basic researches.

澄江古生物研究站
（位于云南澄江，著名的"澄江生物群"发现地）

Chengjiang Field Research Station
(at the discovery site of the "Chengjiang Biota" in Chengjiang, Yunnan)

被誉为"20世纪最惊人的科学发现之一"的"澄江生物群"的发现以及关于"寒武纪大爆发"的研究，引起了国际学术界的轰动。由南京古生物所研究人员与兄弟单位共同主持完成的"澄江生物群与寒武纪大爆发"荣获2003年度国家自然科学奖一等奖。

"金钉子"是地层划分和对比的国际标准或"共同语言"，也是古生物学研究程度和水平在地层划分和对比中的体现。在中国现已建立的10个"金钉子"中，有7个是由南京古生物所科学家主持研究取得的。

南京古生物所有14位科学家先后当选为中国科学院院士（学部委员）:李四光（1951年当选）、斯行健（1951年当选）、赵金科（1980年当选）、卢衍豪（1980年当选）、穆恩之（1980年当选）、王钰（1980年当选）、李星学（1980年当选）、顾知微（1980年当选）、盛金章（1991年当选）、周志炎（1995年当选）、戎嘉余（1997年当选）、金玉玕（2001年当选）、陈旭（2003年当选）、沈树忠（2015年当选）。

Fourteen of NIGPAS scientists have been elected academicians in the Chinese Academy of Sciences: Li Siguang (1951), Si Xingjian (1951), Zhao Jinke (1980), Lu Yanhao (1980), Mu Enzhi (1980), Wang Yu (1980), Li Xingxue (1980), Gu Zhiwei (1980), Sheng Jinzhang (1991), Zhou Zhiyan (1995), Rong Jiayu (1997), Jin Yugan (2001), Chen Xu (2003), Shen Shuzhong (2015).

安徽休宁"蓝田生物群"野外发掘现场

Excavation site of the "Lantian Biota" in Xiuning, Anhui

近年来，南京古生物所与众多国际知名古生物科研机构建立了长期合作关系，主持了若干个国际地质对比计划（IGCP）以及中美、中德、中英等多项重要国际合作项目，主办了"第二届国际古生物学大会"（IPC2006）等一系列具有重要影响力的国际学术会议。研究所科学家在国际古生物协会（IPA）、国际地层委员会（ICS）等诸多国际学术组织中担任了主席、副主席、选举委员等重要职务。

南京古生物研究所拥有一支活跃于国内外学术界的科研队伍、现代化实验技术支持力量、堪称亚洲第一的古生物专业图书馆和历史悠久的古生物标本收藏系统及模式标本库，以及面向公众的古生物博物馆和兼具科研、学术交流和科普功能的澄江古生物研究站。

Many NIGPAS scientists have been actively involved in international academic organizations by serving as officials or leaders, such as in the International Association of Palaeontology, the International Commission of Stratigraphy and its subcommissions, International Organisation of Palaeobotany and International Fossil Specialists Organizations, and the UNESCO's International Geoscience Programmes. NIGPAS has established long-term cooperative relations with many geological and palaeontological institutions worldwide, actively engaging and promoting international collaborations and academic exchanges.

Currently, NIGPAS has diverse research teams in invertebrate palaeontology, palaeobotany and palynology, micropalaeontology, biostratigraphy and chronostratigraphy, with modern laboratories in fossil sectioning, extraction, optic and electronic observation, chemical and molecular analyses; there is a world famous specialized library with collections of geological and palaeontological literature unparalleled in the Asian region and a fossil repository with collections of type fossil specimens dated back to the earliest stage of palaeontology in China. The Nanjing Museum of Palaeontology has been built as a window and interface between the scientists and the public. A field research station has been constructed at the discovery site of the famous "Chengjiang Biota" in Chengjiang, Yunnan with research and conference facilities and a display for the public.

前寒武纪

Precambrian Life Forms

Phanerozoic >

Evolutionary Treasures

迄今为止，地球是唯一已经被发现有生命存在的星球。目前，在这颗蓝色的星球上生活着 870 多万种生物；它们虽然形态各异，但都可以追溯至一个共同的祖先。

据推测，生命大约起源于 38 亿年前。在生命出现后最初的 10 多亿年间，地球上的生命形式一直十分简单，主要为单细胞、原核的细菌和蓝藻等生物。因此，那时的化石记录非常单调，其中绝大多数为叠层石（一种由微生物生长而形成的生物与沉积物复合体）。在距今约 25 亿年的古元古代早期，大气中含氧量显著增加，为真核生物的起源带来了契机。在中元古代（距今 16 亿~10 亿年）早期，原始的单细胞真核生物开始出现，并发生形态分异。进入新元古代（距今 10 亿~5.4 亿年）后，多细胞生物的起源和早期演化是地球生命演化史中的重大革新事件。在距今 7.5~5.8 亿年时，地球上发生了多次全球性的冰期事件，冰层可能覆盖了整个地球，从而形成了所谓的"雪球地球"。冰期结束后，冰川融化导致大规模海侵，多细胞生物的起源和早期演化正是发生在这一特定的地质时期。遍布世界各地的"埃迪卡拉生物群"（最早发现于澳大利亚南部，距今 5.7 亿~5.5 亿年）以及我国发现的"蓝田生物群"（安徽蓝田，距今大约 6 亿年），见证了地球早期生命演化的顶峰阶段。其后是"寒武纪生命大爆发"（开始于距今大约 5.4 亿年）。我们把寒武纪之前这一段漫长而缺少生命的地质时期统称为"前寒武纪"。

As far as we know at the present, the Earth is the only planet in the universe where life forms have been discovered. Estimated some $8.7×10^6$ species currently live on this blue planet, sharing the same atmosphere and hydrosphere, with the human beings as one of them. Although the current life forms are so vastly diverse and abundant, science and natural history suggest that they can trace back to one single origin in deep time.

Researches show that life began about 3.8 billion years ago. For the first billion years after the origin, life forms remained very simple with single prokaryotic cells, most likely belonging to bacteria and archaea. Fossil record of this time is extremely monotonous, most commonly in the form of stromatolites, which are commonly formed by microbial mates trapping and cementing sediment particles. During a time interval beginning about 2.5 billion years ago, referred to as the Palaeoproterozoic Era, due to the significant increase in the atmospheric oxygen, more complex life forms — the eukaryotes — probably started to evolve on the Earth, as indicated by molecular markers from the sediments. However, fossil forms of primitive single-celled eukaryotes did not appear until the Mesoproterozoic Era about 1.6 billion to 1.0 billion years ago, when such eukaryotes also started to differentiate in morphology. The origin and diversification of multicellular organisms were another major evolutionary event of life on Earth, which is uncovered in the fossil record of the Neoproterozoic Era about 1.0 billion years to 540 Ma. During the Neoproterozoic, intensive global glaciations occurred a number of times from about 750 Ma to 580 Ma, the so-called "Snowball Earth" event. Notably, the evolution of multicellularity on Earth must be closely related to the "Snowball Earth" event and post-glaciation environmental changes. The globally occurring Ediacaran Biota (first discovered in southern Australia; about 570–550 Ma) and the newly found Lantian Biota (Lantian, Anhui, China; about 600 Ma) marked the climax of early life evolution with complex life forms before the Cambrian Explosion which started about 540 Ma. Thus, all the geological time covered in this chapter is referred to as "Precambrian".

Evolutionary Treasures

在中国，前寒武纪地层记录非常丰富，前寒武纪古老地层保存了大量 38 亿年前到 5.4 亿年前的沉积岩石和化石，成为探索地球早期环境和早期生命演化的窗口。

前寒武纪叠层石和藻类等化石的研究在我国取得重要进展。南京古生物所几代学者长期致力于最古老的生物化石的发现和研究，致力于破译早期生命起源和进化的密码，在中元古代真核生物化石、"淮南生物群"、"蓝田生物群"、"瓮安生物群"、"庙河生物群"、"高家山生物群"，以及典型的"埃迪卡拉生物群"研究中取得突破。他们发现了已知最早的、距今 6 亿年的地衣化石，研究了地球上迄今发现最早的、距今 6.35 亿 ~5.8 亿年的复杂宏体生物组合——"蓝田生物群"的地球生命演化，并在"瓮安生物群"中发现了多样化复杂后生生物化石。这一系列成果标志着我国地球生命演化早期的研究已走在国际学术前沿。

China is rich in Precambrian stratigraphic records, where tremendously abundant sedimentary rocks and fossils are preserved dating from 3.8 billions years ago to 541 Ma. These excellent geological records serve as windows for exploring the early terrestrial environment and early life evolution on the Earth.

Significant advances in studies of Precambrian fossils, including stromatolites, algae and others, have been made in China. Generations of NIGPAS scholars have been vigorously exploring for the oldest fossils and different life forms in the deepest geological records, in order to understand the evolution of early life on the Earth. Breakthroughs have been made in the discovery and interpretation of the Mesoproterozoic eukaryotes, the Huainan Biota, Lantian Biota, Weng'an Biota, Miaohe Biota, Gaojiashan Biota and typical Ediacaran fossils in the Yangtze Gorges area. The findings of the oldest lichen fossils (about 600 Ma) from the Weng'an Biota and the complex metazoan-type macrofossil assemblages from the Lantian Biota (about 635–580 Ma) are considered academic frontiers in early life researches.

Archean >   Proterozoic >

4000   2500   1600   1000   541.0±1.0
Time (Ma)

Stromatolite
叠层石
中元古代（距今约 14 亿年），天津蓟县

Mesoproterozoic; Jixian, Tianjin
About 1.4 billion years ago

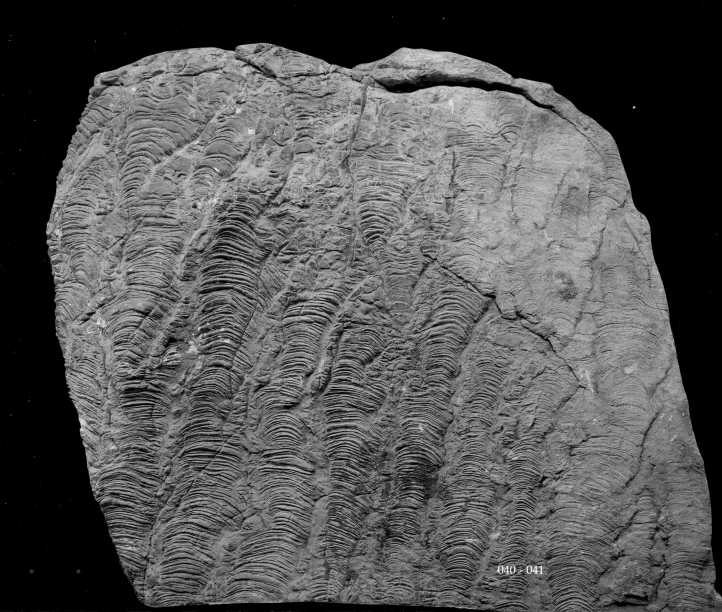

Evolutionary Treasures

叠层石 / Stromatolites

叠层石是以蓝细菌（蓝绿藻）为主的微生物席，是在生长和新陈代谢活动过程中通过黏附、沉淀矿物质或捕获矿物颗粒而形成的一种生物沉积构造。形成叠层石的微生物群落最早出现于大约 35 亿年前。当时的地球环境异常恶劣，这些微生物群落的构成可能相当复杂，包括多种多样的代谢机理。它们对环境资源的竞争以及它们的差异性死亡等生命活动，造就了我们所见到的叠层石的精细结构。大型叠层石在前寒武纪海洋中比较常见，是确定前寒武纪地层时代和进行地层对比的理想工具。叠层石在显生宙和现今海洋中则非常稀少，这是由全球环境的巨大变化，特别是底质生物群落中宏体生物扰动的出现所造成。因此，显生宙以后只有在极端条件下（例如高温热泉、高盐海湾等不适合宏体生物在底质中生长的环境，就像今天的巴哈马群岛、西澳大利亚的鲨鱼湾和美国的黄石公园）才可能形成微生物主导的叠层石。

Stromatolites are rock-like buildups of microbial mats, mainly consisting of photosynthetic/chemosynthetic algae and bacteria, with laminated accretionary structures formed by baffling, trapping, and precipitation of particles by communities of microorganisms in limestone- or dolostone-sedimentary environment. Stromatolite-building microbial communities include the oldest known fossils, dating back some 3.5 billion years when the environments of the Earth were too hostile to support life as we know it today. We can presume that the microbial communities consisted of complex consortia of species with diverse metabolic needs, and that competition for resources and differing motility among them created the intricate structures we observe in these ancient fossils.

Large stromatolites were common in the Precambrian oceans on the Earth, becoming the best stratigraphic tool for dating and correlation of sedimentary rocks around the world; but they are rare today, only found in extreme environment (e.g., high salinity sea water, hot springs, etc.) where macro-organisms are unable to live in the substrate to disturb the continuous growth of microbial mats, such as in the Bahamas, the Shark Bay of western Australia and the Yellowstone National Park.

Archean >   Proterozoic >

4000   2500   1600   1000   541.0 ± 1.0
Time (Ma)

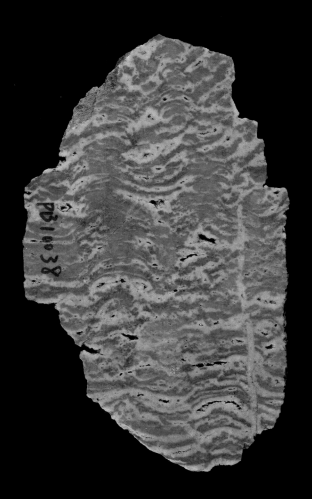

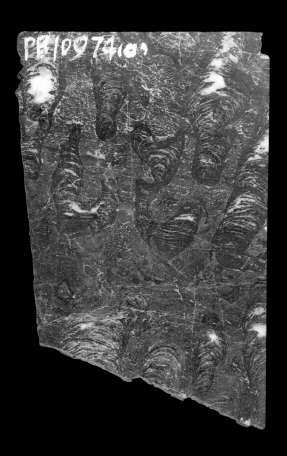

Gymnosolen crass
厚壁裸枝叠层石
新元古代，新疆尉犁库鲁克塔格

Neoproterozoic; Kuruktag, Yuli, Xinjiang
NIGPAS Collection #PB10974

Pratum junctus
连接草地叠层石
新元古代，内蒙古额济纳旗

Neoproterozoic; Ejina Qi, Inner Mongolia
NIGPAS Collection #PB10038

Evolutionary Treasures

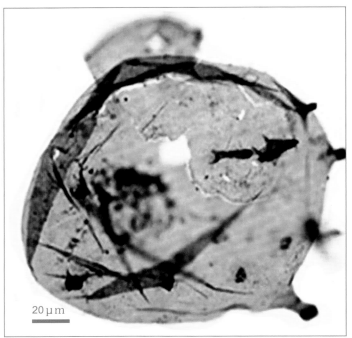
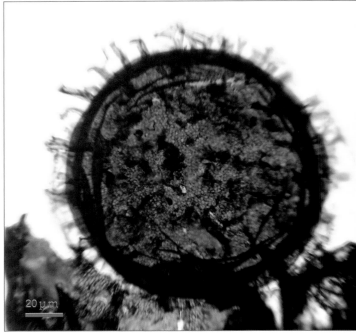

中元古代疑源类化石（距今约13亿年），山西永济

左图(left): *Tappania plana* 平坦塔潘藻
右图(right): *Shuiyousphaeridium macroreticulatum* 大网水幽藻

Mesoproterozoic; Yongji, Shanxi; about 1.3 billion years ago

疑源类　/　Acritarchs

疑源类是具有机壁的、亲缘关系不明的微体浮游生物类群，可能包括一些后生动物的卵鞘、藻类的休眠囊等。疑源类化石最早出现于太古宙地层中，在前寒武纪地层的研究中发挥了重要作用。疑源类在显生宙继续存在。疑源类化石不仅是前寒武纪和古生代地层划分对比的重要依据，也是地球早期生命演化和地球生态系统演变的重要证据。

新元古代疑源类（距今约7亿年），安徽淮南

左图(left): *Pololeptus rugosus*
　　　　　褶皱端弱球藻
右图(right): *Trachyhystrichosphaera aimika*
　　　　　埃米卡拟粗面刺球藻

Neoproterozoic; Huainan, Anhui; about 700 Ma

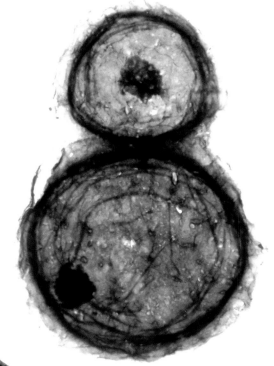

Acritarchs are organic-walled, phylogenetically uncertain microfossils, probably including egg cases of some small metazoans and resting cysts of some algae. Fossil record of acritarchs can date back to the Archean Eon to the present; these fossils are not only very important for stratigraphic division and global correlation, but also especially significant for understanding evolution of Precambrian life and ecosystem changes.

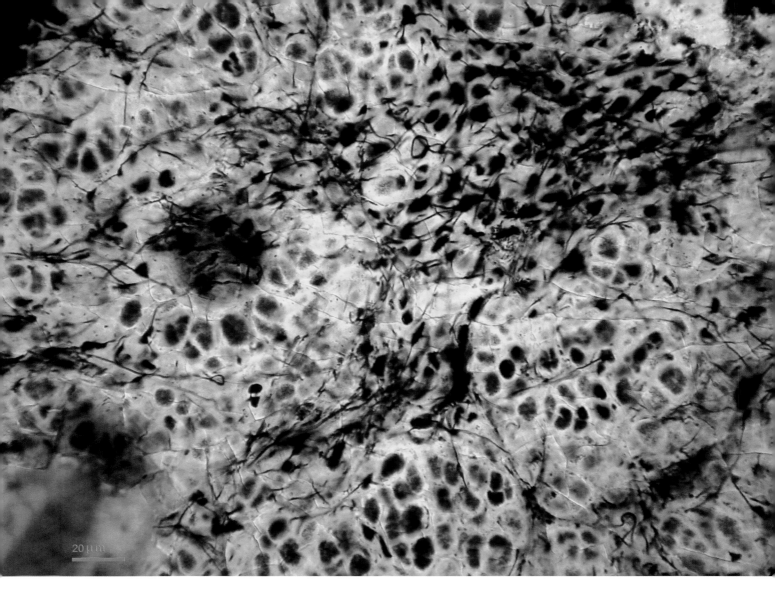

最古老的地衣（距今约6亿年，来自贵州瓮安）

The oldest fossil lichen from Weng'an, Guizhou; about 600 Ma (published in *Science*, vol. 308, 2005)

"瓮安生物群" / Weng'an Biota

距今约 6 亿年的"瓮安生物群"产于贵州瓮安磷矿埃迪卡拉系陡山沱组磷块岩中，它以保存了磷酸盐化的精美细胞结构而闻名于世，是认识多细胞生物早期演化的一个重要窗口。"瓮安生物群"中已经发现的化石类群包括动物胚胎、动物休眠卵、具有营养细胞和繁殖细胞的分化多细胞真核生物等化石。相关的一系列重要成果已在 *Science*、*Nature*、*PNAS* 等国际著名刊物上发表。

Evolutionary Treasures

Eocyathispongia qiania
贵州始杯海绵
瓮安生物群（距今约6亿年），贵州

Weng'an Biota from Guizhou; about 600 Ma
(published in *PNAS*, vol. 112, 2015)

The Weng'an Biota is a well-known Late Neoproterozoic biota (about 600 Ma) of the Ediacaran Period with exceptional preservation of delicate phosphotized cell structures, discovered from the phosphorite deposits of the Doushantuo Formation at the Phosphorite Mine in Weng'an, Guizhou. The Weng'an Biota includes exceptionally preserved eukaryote fossils, including diverse acanthomorphic acritarchs, tubular and spheroidal microfossils, many of which have been interpreted as multicellular eukaryotes, including sponges, animal embryos and resting eggs, and differentiated vegetative and reproductive tissues of multicellular organisms, as published in leading academic journals such as *Nature*, *Science* and *PNAS*.

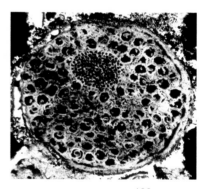

显示细胞分化（营养细胞和繁殖细胞）的早期多细胞真核生物
瓮安生物群（距今约6亿年），贵州

Late Neoproterozoic multicellular organism showing cell differentiation and germ-soma separation
Weng'an Biota from Guizhou; about 600 Ma (published in *Nature*, vol. 515, 2014)

新元古代蓝田生物群复杂宏体化石（距今约6亿年，安徽蓝田）

Doushantuophyton cometa
帚状陡山沱藻

Neoproterozoic complex macrofossils from Lantian Biota, southern Anhui; about 600 Ma (published in *Nature*, vol. 470, 2011)

5mm

"蓝田生物群" / Lantian Biota

在现今生物圈中，所有肉眼可见的生命几乎都是多细胞宏体生物，即人们常说的"高等生命"。多细胞宏体生物的出现是地球生命进化史上极为重要的革新事件。产自安徽休宁的距今6.3亿~5.8亿年的"蓝田生物群"，被认为是迄今最古老的宏体真核生物群化石，不仅为多细胞生物的起源提供了更古老的化石证据，也指示了当时大气圈和浅层海水中的氧气含量足以支持多细胞生物的生存和发展。《自然》杂志评述指出："'蓝田生物群'为早期复杂宏体生命的研究打开了一个新窗口"。

Archean >   Proterozoic >

Evolutionary Treasures

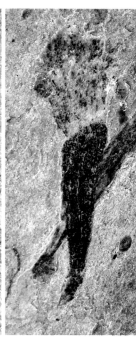

5 mm

In the present biosphere, almost all visible life forms (macro-organisms) are multicellular organisms, often referred to as the higher organisms, which are less known before the Cambrian Explosion. The emergence of multicellular complex and differentiated organisms during the Precambrian is a major evolutionary innovation event in the history of life on the Earth, categorically parallel to other major events such as the origin of eukaryotes and the Cambrian metazoan radiation. The Lantian Biota, discovered from Precambrian strata about 635–580 Ma in Xiuning, Anhui, represents the oldest multicellular eukaryotic macrofossil assemblages, so far as known, which is said to have opened a new window for the evolutionary study of early complex life forms, in a commentary published by the journal *Nature*. The Lantian Biota contains a diverse group of macrofossils including animal-like organisms and multicellular algae, which may indicate that the oxygen content of the atmosphere and at least the upper layer of the ocean during that period was high enough to support the growth and survival of multicellular macro-organisms.

左图(left): *Orbisiana linearis*
线状奥尔贝串环

右图(right): *Lantianella laevis*
光滑蓝田虫

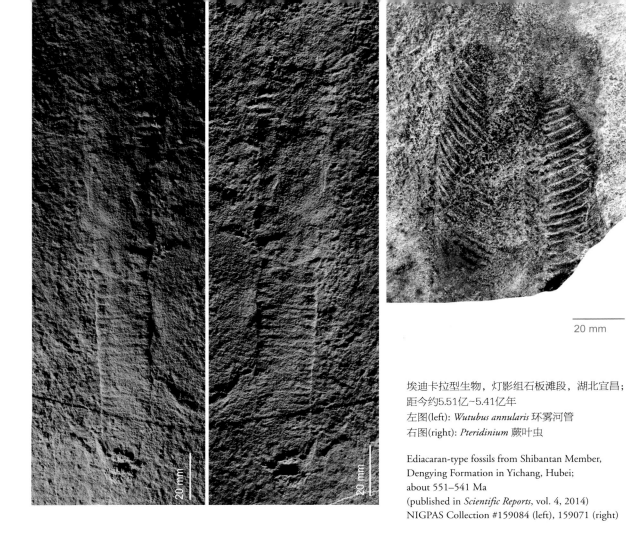

埃迪卡拉型生物，灯影组石板滩段，湖北宜昌；
距今约5.51亿~5.41亿年
左图(left): *Wutubus annularis* 环雾河管
右图(right): *Pteridinium* 蕨叶虫

Ediacaran-type fossils from Shibantan Member,
Dengying Formation in Yichang, Hubei;
about 551–541 Ma
(published in *Scientific Reports*, vol. 4, 2014)
NIGPAS Collection #159084 (left), 159071 (right)

"埃迪卡拉型生物群" / Ediacaran-type Biota

典型的埃迪卡拉生物是一些管状、蕨叶状、多数固着生活的宏体生物，因最早发现于澳大利亚埃迪卡拉山而得名。它们是"寒武纪大爆发"前夕最为独特的宏体生物群，出现于世界各地新元古代埃迪卡拉纪（6.35亿~5.41亿年前）（曾称"文德纪"，中国称"震旦纪"），被认为是生物演化的一次失败的尝试，在"寒武纪大爆发"之前几乎灭绝（极少数残存到寒武纪早期）。长期以来，这类化石在中国未被发现。近年来，在三峡地区灯影组碳酸盐岩中发现的多种类型的埃迪卡拉型生物群典型分子，为探索这类奇特生物群的取食方式和生态空间等重要科学问题打开了一扇新窗口。在中国，典型的埃迪卡拉型生物群出现于蓝田生物群和瓮安生物群之后。

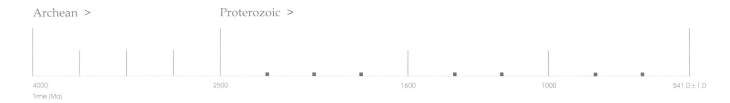

Evolutionary Treasures

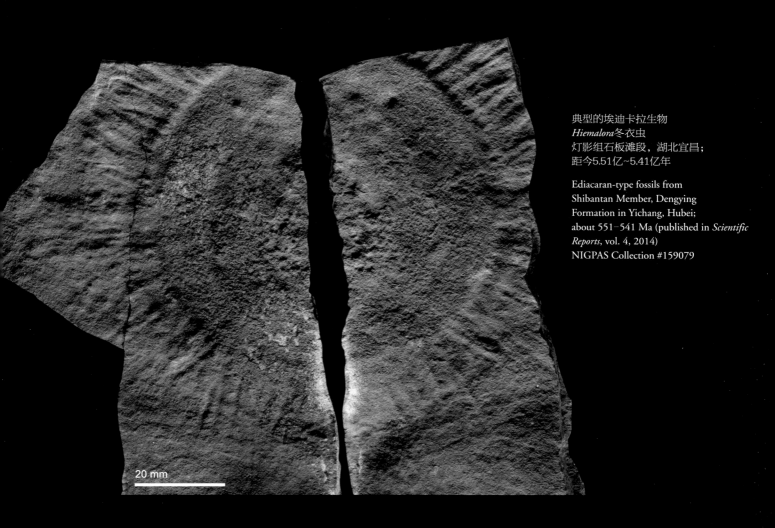

典型的埃迪卡拉生物
Hiemalora 冬衣虫
灯影组石板滩段，湖北宜昌；
距今5.51亿~5.41亿年

Ediacaran-type fossils from
Shibantan Member, Dengying
Formation in Yichang, Hubei;
about 551–541 Ma (published in *Scientific Reports*, vol. 4, 2014)
NIGPAS Collection #159079

Typical Ediacaran Biota consists of frond-like and tubular macrofossils of mostly sessile organisms, first discovered in the Ediacara Hills, South Australia. They occur worldwide in rocks of the Neoproterozoic Ediacaran Period (635—541 Ma), previously referred to as the "Vendian Period" or the "Sinian Period" in China. Due to its peculiarity in morphology and the ecology, this biota is considered to represent "a failed evolutionary experiment" that was probably completely replaced by the Cambrian biotas including animal body plans that we can recognize all through the Phanerozoic. Interesting, the typical Ediacaran Biota had not been discovered in China until very recently when NIGPAS scientists excavated multiple Ediacaran-type fossils in the Yangtze Three Gorges region, a significant advance in this research direction which is likely to provide a new window for uncoding the systematics and ecology of the Ediacaran Biota. In China, these Ediacaran-type fossils are younger than the Lantian Biota and the Weng'an Biota.

Archean >   Proterozoic >

Time (Ma)
4000 3600 3200 2800 2500 2300 2050 1800 1600 1400 1200 1000 720

古生代

Palaeozoic Animals and Plants

Phanerozoic >

Evolutionary Treasures

随着埃迪卡拉型生物群的神秘消失，地球在 5.41 亿年前迈入了一个新的时代——古生代（显生宙）。从此，生命世界变得丰富多彩起来。从 5.41 亿年前到 2.52 亿年前，这段长达近 3 亿年的地质年代被称为古生代。古生代分为 6 个纪：寒武纪、奥陶纪、志留纪、泥盆纪、石炭纪和二叠纪。古生代生命演化经历了寒武纪生命大爆发、奥陶纪生命大辐射、志留纪—泥盆纪鱼类的繁盛和陆生维管植物的辐射、石炭纪两栖动物和森林的繁盛等重大发展阶段，同时也经历了奥陶纪末、泥盆纪中后期以及二叠纪末三次生物大灭绝事件。

With the mysterious disappearance of the Ediacaran Biota during latest Neoproterozoic, the Earth entered a completely new era — Palaeozoic about 541 Ma, when life began to flourish first in the ocean and then on land. Palaeozoic — the first era of the Phanerozoic Eon, spans the period from 541 Ma to 252 Ma, including the periods of Cambrian, Ordovician, Silurian, Devonian, Carboniferous and Permian. The Palaeozoic biotic world experienced dramatic evolutionary events such as the Cambrian Explosion, the Great Ordovician Radiolarian and end-Ordovician Extinction, the Silurian–Devonian radiations of vertebrates and vascular plants (probably the major terrestrialization event), Frasnian-Famenian (F-F) mass extinction, the Carboniferous flourishing of amphibians and forests on land, and the end-Permian mass extinction.

"澄江生物群"产出地层，寒武系第二统筇竹寺组玉案山段，云南澄江

The rocks yielding the "Chengjiang Biota", Yu'anshan Member of the Qiongzhusi Formation, Cambrian 2nd Series; Chengjiang, Yunnan

Phanerozoic >

Ediacaran	Cambrian	Ordovician	Silurian	Devonian	Carboniferous	Permian	Triassic	Jurassic	Cretaceous	Paleogene	Neogene	Quaternary
	Palaeozoic						Mesozoic			Cenozoic		
541.0 ± 1.0	485.4 ± 1.9	443.8 ± 1.5	419.2 ± 3.2	358.9 ± 0.4	298.9 ± 0.15	251.902 ± 0.024	201.3 ± 0.2	145	66	23.03	2.58	Present

Evolutionary Treasures

"澄江生物群"与"寒武纪大爆发"

1984 年,"澄江生物群"被南京古生物所的科研人员首先发现于云南澄江帽天山距今约 5.2 亿年的寒武纪早期地层中(筇竹寺组帽天山页岩),是著名的寒武纪特异保存化石群。它以保存大量软躯体宏体化石为特征,被认为是寒武纪生命"大爆发"的窗口,比其前发现于加拿大的著名的"布尔杰斯生物群"早 1000 多万年(寒武纪中期)。"澄江生物群"含有丰富的动物和藻类化石,已正式报道动物化石 229 种,涵盖现今地球上几乎所有的动物门类的化石(仅苔藓动物化石未曾发现于这个时期)。由中科院南京古生物所、云南大学和西北大学科学家合作的"澄江生物群"研究成果荣获 2003 年度国家自然科学奖一等奖。

"寒武纪大爆发",也称"寒武纪大辐射",是地球生命史上最壮观的一次生物演化事件;宏体生物,尤其是具有矿化骨骼的生物开始登上地球生命的舞台,并成为压倒性的主角。寒武纪(显生宙古生代开始)始于 5.41 亿年前,此后的大约 2000 万年间,地球上快速演化出现几乎所有动物门类。在寒武纪之前,地球上的生物基本上都是单细胞的简单生物,有些形成细胞群集,而复杂宏体生物极为少见。"寒武纪大爆发"彻底改变了地球生态面貌,海洋中出现了大量多样化的宏体生物,并出现大规模矿化骨骼,底栖、漂浮、游泳以及捕食等各种生态方式也快速涌现。地球表面由微小简单生物主导的生态面貌,经过漫长的演化历程(大约 30 亿年)后发生突变,从此,复杂的地球生态系统的演化进程开启了。自达尔文以来,"寒武纪大爆发"一直被认为是演化生物学中的一个谜。为何此时?为何如此突然?是否真实?这些问题仍是众说纷纭。尽管从地质学、古生物学和演化生物学角度提供了许多解释,但"寒武纪大爆发"的机制和演化速率等问题仍然是科学界须继续探究的科学难题。

The "Chengjiang Biota" and the "Cambrian Explosion"

The "Chengjiang Biota", a world-famous fossil Lagerstätten designated a World Natural Heritage, was first discovered by NIGPAS researchers in 1984, from the Maotianshan Shale (Yu'anshan Member, Qiongzhusi Formation) of the Early Cambrian age approximately dated back to 520 Ma, at Mt. Maotianshan, Chengjiang, Yunnan. It is characterized by abundant and diverse macrofossil forms with many soft-body structures exceptionally preserved. The Chengjiang Biota is a window of the Cambrian Explosion, more than 10 million years older than the well-known Middle Cambrian Burgess Biota of British Columbia, Canada. The Chengjiang Biota includes more than 229 species of nearly all animal phyla except Bryozoa which have not been reported from the Cambrian strata. Research achievements by scientists from NIGPAS, Yunnan University and Northwest University (Xi'an) together have been awarded the first prize for National Natural Science Award in 2003 in China.

The "Cambrian Explosion", or the "Cambrian Radiation", occurred during the Early Cambrian (about 541 Ma), after experienced a short period of small shelly faunas, is the most magnificent evolutionary event, especially for the macro-animals with biomineralized skeletons who suddenly (i.e., in a geological instant) become the ruling parties in the Earth ecosystems. During the initial 20 million years or so of the Cambrian Period (Phanerozoic Eon), there evolved nearly all animal phyla (except Bryozoa) apparently from nothing, because all we know from earlier fossil record are simple single-cell organisms or their colonies while complex macrofossils are extremely scarce. The Cambrian Explosion dramatically changed the Earth biosphere from a microbial-dominating world to an ecologically and morphologically diverse macro-organic world, where the oceans were homes of swimming, floating, bottom-dwelling animals forming complex ecosystems, in addition to the evolving micro-organic world; the latter dominated the Earth surface for some 3 billion years. The Cambrian Explosion has been puzzling to scientists since Darwin as to questions such as "Why this time? Why so abrupt? Or is it real? ", to which various different interpretations have been put forward and still remain frontal issues in evolutionary biology and palaeontology.

Evolutionary Treasures

纳罗虫 / *Naraoia*

纳罗虫（*Naraoia*）是一类具有传奇色彩的特殊三叶虫，"澄江生物群"及其他寒武纪爆发代表生物群中最常见的节肢动物之一。1984 年，南京古生物研究所硕士研究生侯先光前往云南澄江帽天山进行野外采集时，在寒武系第二统筇竹寺组玉案山段下部发现了一个保存软体和精美附肢等形态构造的多门类无脊椎动物化石群。1985 年 11 月《古生物学报》刊登了"*Naraoia* 在亚洲大陆的发现"一文，揭开了"澄江生物群"研究的第一幕。

Naraoia is a legendary genus of special trilobites, one of the most common arthropods in the "Chengjiang Biota" and other faunas of the Cambrian Explosion event. In 1984, when a graduate student at NIGPAS, Hou Xianguang went to Mt. Maotianshan, Chengjiang of Yunnan to do field collection, he found a diverse fauna with exceptionally preserved soft-body parts and appendages. In November 1985, a research paper, entitled "Preliminary notes on the occurrence of the unusual trilobite *Naraoia* in Asia", appeared in *Acta Palaeontologica Sinica*, marking the beginning of a research spree for the "Chengjiang Biota".

Naraoia spinosa
刺状纳罗虫
南京古生物所发现的第一批澄江生物群化石
寒武系第二统筇竹寺组玉案山段，云南澄江
The first specimen collected from the Chengjiang Biota by NIGPAS
Yu'anshan Member of the Qiongzhusi Formation, Cambrian 2nd Series; Chengjiang, Yunnan

Evolutionary Treasures

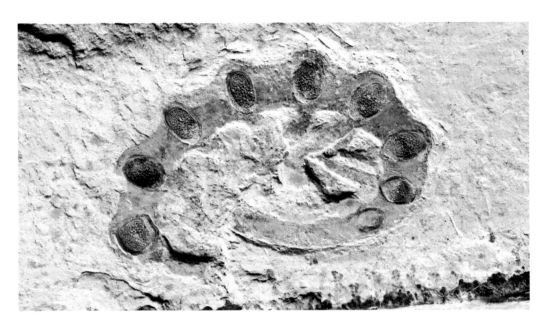

Microdictyon
微网虫
寒武系第二统筇竹寺组玉案山段，云南澄江

Yu'anshan Member of the Qiongzhusi Formation, Cambrian 2nd Series; Chengjiang, Yunnan

微网虫 / *Microdictyon*

微网虫属于叶足动物门，因身上多边形的鳞状骨片而得名，被称为"带盔甲的蠕虫"，其化石发现于"澄江生物群"中。在此之前，研究者仅发现了分散保存的骨片，而无从知道其动物原型。微网虫身上的鳞片状骨片可能是具有视觉功能的构造，相应部分伸出触角状足。它曾刊登在《自然》杂志封面上，是"寒武纪大爆发"中的一个"明星"。

Microdictyon belongs to a group of worm-like animals with stubby legs named lobopodians. It has net-like skeleritic scales with tentacle feet below, known from the "Chengjiang Biota"; earlier researchers only found separate scales from sediments, unable to reconstruct the original animal. A photo of *Microdictyon* was published as a cover photo in *Nature* magazine, becoming one of the "fossil stars" of the Cambrian Explosion event.

Microdictyon
微网虫
寒武系第二统筇竹寺组玉案山段，云南澄江

Yu'anshan Member of the Qiongzhusi Formation, Cambrian 2nd Series; Chengjiang, Yunnan

Evolutionary Treasures

Anomalocaris
奇虾
寒武系第二统筇竹寺组玉案山段,云南澄江

Yu'anshan Member of the Qiongzhusi Formation, Cambrian 2nd Series; Chengjiang, Yunnan

奇虾 / *Anomalocaris*

奇虾是寒武纪海洋的大型掠食者,位于食物链金字塔的顶端。据估计,奇虾体长可能超过 2 m,以三叶虫等为食。奇虾类化石在中国寒武纪特异埋藏动物化石群中有广泛报道,包括"澄江生物群"、"牛蹄塘动物群"、"关山动物群"、"马龙动物群"和"凯里动物群"等。近年来在河北地区寒武系馒头组也发现了奇虾化石。

Anomalocaris is a giant predator of the Cambrian ocean, occupying the top niche in the pyramidal food chain. It is estimated that *Anomalocaris* could be as large as over 2 m in length, feeding on trilobites and other smaller animals. It has been widely reported from a number of Chinese Cambrian Lagerstätte faunas, including the Chengjiang, the Niutitang, the Guanshan, the Malong, and Kaili biotas in Yunnan and Guizhou provinces, as well as a newly discovered one from the Cambrian Mantou Formation in Hebei Province.

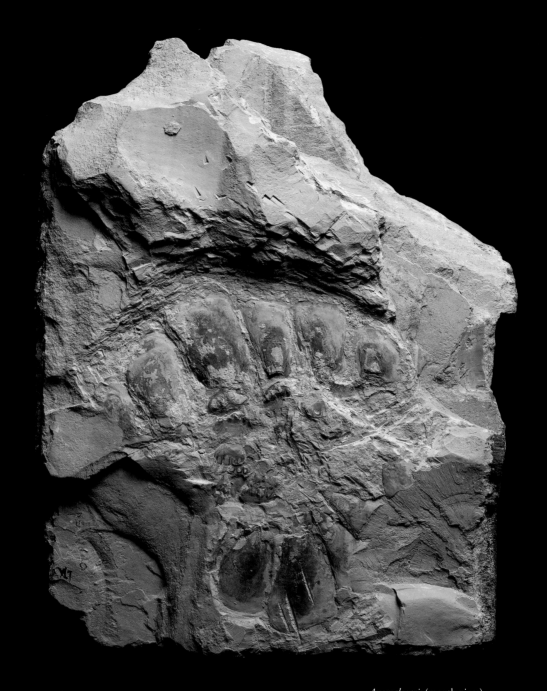

Anomalocaris (mouth piece)
奇虾类口器
寒武系第二统筇竹寺组玉案山段,云南澄江

Yu'anshan Member of the Qiongzhusi
Formation, Cambrian 2nd Series;
Chengjiang, Yunnan

Fuxianhuia protensa
延长抚仙湖虫
寒武系第二统筇竹寺组玉案山段，云南澄江

Yu'anshan Member of the Qiongzhusi Formation,
Cambrian 2nd Series; Chengjiang, Yunnan

抚仙湖虫 / *Fuxianhuia*

抚仙湖虫化石是"澄江生物群"中特有的化石。抚仙湖虫的头部分节特征和躯干附肢形态代表了真节肢动物中的原始类型。抚仙湖虫具备复杂的脑结构（脑部两侧具有三组视觉纤维网），这与昆虫类和软甲类生物相仿。

Fuxianhuia is a peculiar arthropod of the Chengjiang Biota, named after the Fuxian Lake near its discovery site. The genus is regarded as a primitive type of euarthropods due to the features in its cephalic segmentation and trunk appendages. In addition, *Fuxianhuia* shows the earliest complex brain structure with 3 optic neuropils, indicating its resemblance to insects and malacostracans.

Fuxianhuia protensa
延长抚仙湖虫
寒武系第二统筇竹寺组玉案山段，
云南澄江

Yu'anshan Member of the Qiongzhusi Formation, Cambrian 2nd Series; Chengjiang, Yunnan

海口虫 / *Haikouella*

海口虫是澄江生物群的特殊动物，属于脊索动物门有头类。它的形态为扁平状，似文昌鱼；体长一般为 20~30 mm，具有头、鳃、脑、咽齿、脊索、心脏和循环系统、肌节和消化道等解剖特征。因此，它被认为是脊椎动物的祖先类群。

Haikouella is a group of particular animals from the Chengjiang Biota, belonging to chordates and likely craniates. It's body is flat like lancelet, normally 20–30 mm in length, with a head, gills, brain, pharyngeal teeth, a notochord, a heart and the circulatory system, segmental blocks of muscle fibers (myomeres), and zhe gut with esophagus, spiral midgut and straight intestine, anatomical features considered to indicate its ancestral relationship to vertebrates.

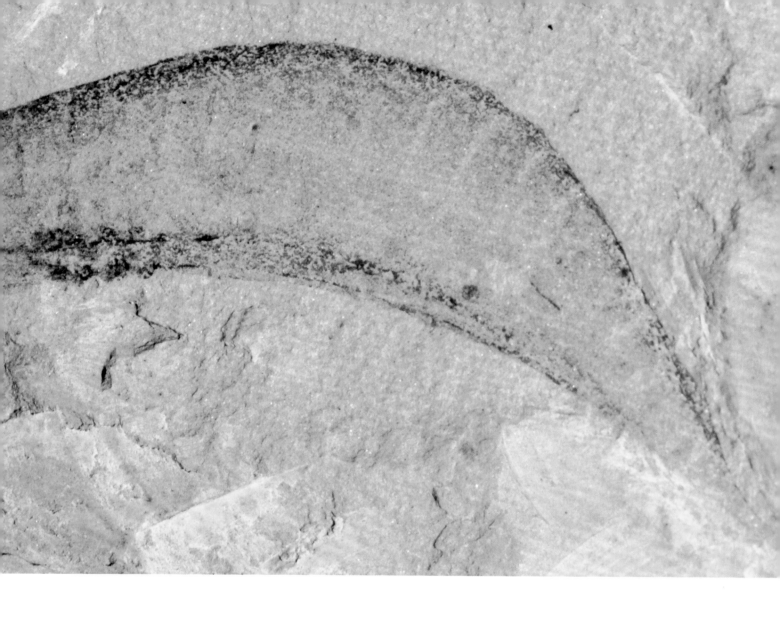

Haikouella lanceolata
梭形海口虫
寒武系第二统筇竹寺组玉案山段，云南澄江

Yu'anshan Member of the Qiongzhusi
Formation, Cambrian 2nd Series; Chengjiang, Yunnan
NIGPAS Collection #164123a

海口虫复原图
Cartoon of *Haikouella*

Evolutionary Treasures

Jianfengia multisegmentalis
多节尖峰虫
寒武系第二统筇竹寺组，云南澄江

Qiongzhusi Formation, Cambrian 2nd Series; Chengjiang, Yunnan

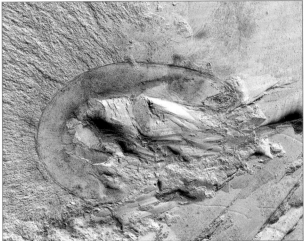

Paucipodia inermis
无饰贫腿虫
寒武系第二统筇竹寺组，云南澄江

Qiongzhusi Formation, Cambrian 2nd Series; Chengjiang, Yunnan

Burithes yunnanensis
云南薄氏螺
寒武系第二统筇竹寺组，云南澄江

Qiongzhusi Formation, Cambrian 2nd Series; Chengjiang, Yunnan

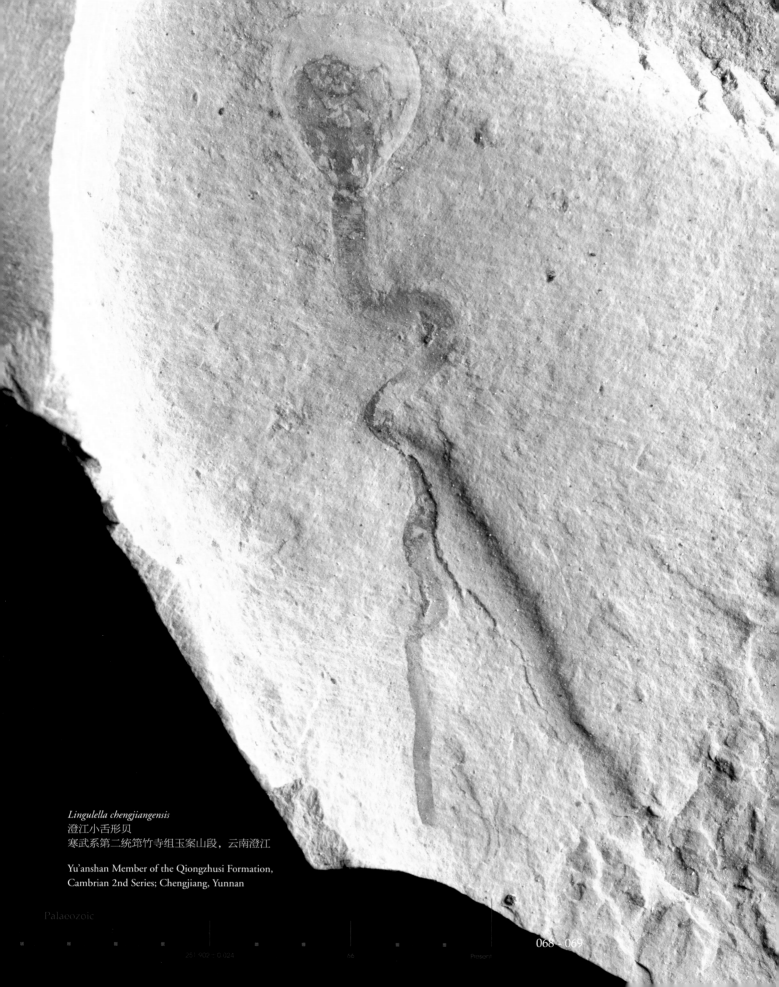

Lingulella chengjiangensis
澄江小舌形贝
寒武系第二统筇竹寺组玉案山段，云南澄江

Yu'anshan Member of the Qiongzhusi Formation,
Cambrian 2nd Series; Chengjiang, Yunnan

南京古生物博物馆一角

A view of exhibits in the Nanjing Museum of Palaeontology

早古生代生命 / Early Palaeozoic Life

早古生代海洋见证了各类动物的起源、演化和复杂生态系统的形成。寒武纪初期以澄江生物群为代表的生物大爆发拉开了显生宙波澜壮阔的生物演化历史。三叶虫、海绵动物、鹦鹉螺、笔石、腕足动物、珊瑚、牙形动物、苔藓动物等在早古生代海洋中轮番登场。奥陶纪生物大辐射为古生代海洋生物群的发展确立了总体构架，不同门类的生物适应不同的生活环境，逐渐发生生态分异，有的营底栖固着、移动或底质内生活，有的营游泳或漂浮生活。奥陶纪晚期的大冰期伴随显生宙第一次生物大灭绝，致使大约一半的海洋生物消失。无颌类脊椎动物（包括甲胄鱼和牙形类）出现于早古生代早期海洋，到奥陶纪和志留纪时已演化出适应淡水环境的脊椎动物和无脊椎动物。晚志留世出现的有颌类脊椎动物和早期维管植物代表生物演化史上的重大事件——动植物登陆，说明早古生代晚期的地球生态系统已经从海洋入侵陆地，使得陆地逐渐摆脱了荒无生机的景象。这充分表明，生命演化不仅受到地球环境演变的驱使，而且驱动地球环境发生了深刻的变化。

Early Palaeozoic seas and oceans witnessed dramatic changes in animal evolution and the formation of complex ecosystems on the Earth. Major animal groups such as trilobites, sponges, brachiopods, graptolites, corals, conodonts, and bryozoans originated and successively populated the Early Palaeozoic marine environment. After the tremendous Cambrian Explosion, there occurred the second major animal radiation during the Palaeozoic — the Great Ordovician Biodiversification Event (GOBE) which transformed the Cambrian Fauna into the Palaeozoic Fauna, involving significant ecological differentiation of different life styles such as the bottom dwellers of sessile, free locomotion and infaunal types, the swimmers at different levels of the water column and the planktons. The end Ordovician mass extinction event — the first of Phanerozoic — was related with a global glaciation event, killing nearly half of all marine animal genera. Agnathan vertebrates (conodonts and ostracoderms) evolutioanarily appeared in the Early Palaeozoic oceans, while freshwater invertebrates and vertebrate animals appeared during Ordovician — Silurian time. By Late Silurian, fossil record shows the appearance of gnathan vertebrates and vascular plants, both would later massively colonize the land surface — another major evolutionary event of the Earth life, that changed the terrestrial environment from a harsh lifeless rocky surface to a green amiable scenario. The evolutionary history interestingly shows that the evolution of life not only is driven by changes in the physical environment, but also causes gradual changes in the Earth environment.

Evolutionary Treasures

三叶虫 / Trilobites

寒武纪被称为"三叶虫时代"。三叶虫是寒武纪海洋的主宰，属于节肢动物门。三叶虫的身体分节，分为头、胸、尾三部分，同时有两条带沟将身体分为垂直三叶，故此得名。它们最早出现于寒武纪早期，并于早古生代达到顶峰，种类极为丰富，为全球性分布种。三叶虫在古生代晚期逐渐减少，最后在二叠纪末生物大灭绝中消失。三叶虫在生物地层学、演化生物学、古生物系统学、古地理学、板块构造学等领域，具有重要的研究意义。

The Cambrian Period is known as "the age of trilobites", due to trilobites' dominance in the Cambrian marine faunas. Trilobites belong to Arthropoda, characterized by a tripartite body composed of the head, the trunk and the tail, and also three vertical "lobes" (the name "trilobites" thus derived) separated by two longitudinal furrows. Trilobites first appeared during early Cambrian, reached their developmental acme during the Early Palaeozoic (Cambrian, Ordovician and Silurian) with high diversity and global distribution, and gradually declined afterwards, until the end Permian mass extinction event when they finally disappeared from the fossil record. Studies of trilobites contribute significantly to biostratigraphy, evolutionary biology, palaeontology, palaeogeography and plate tectonics.

Lotagnostus asiaticus
亚洲花球接子
NIGPAS Collection #7748

Westergaardites pelturaeformis
小尾状韦氏虫
NIGPAS Collection #7851
寒武纪，新疆西库鲁克塔格

Cambrian; Western Kuruktag, Xinjiang

Eoredlichia intermedia
中间型古莱得利基虫
寒武系第二统，云南澄江

Cambrian 2nd Series; Chengjiang, Yunnan

Palaeozoic

Asaphopsoides yongshunensis
永顺似栉壳虫
早奥陶世，湖南永顺

Early Ordovician; Yongshun, Hunan
NIGPAS Collection #MI121

Evolutionary Treasures

Paraszechuanella sp.
副四川虫（未定种）
早奥陶世，湖南

Early Ordovician; Hunan
NIGPAS Collection #MI28

中国的三叶虫化石研究始于20世纪20年代，是国人系统研究化石最早发端的门类之一。自孙云铸教授1924年出版第一本专著以后，三叶虫化石研究陆续取得很大进展，且人才辈出，孙云铸、盛莘夫、卢衍豪、张文堂等成为中国三叶虫学科的奠基者。南京古生物所历经几代人的努力，在三叶虫系统分类、生物地层、古生物地理区系和个体发育等方面长期瞄准国际前沿。近年来，在寒武系全球年代地层学领域引领国际前沿。

Studies of trilobites in China began in the 1920s, as one of the first subdisciplines of palaeontology. Since then, a number of generations of scholars in China have progressively advanced researches on trilobites; representative researchers such as Sun Yunzhu, Sheng Xinfu, Lu Yanhao, and Zhang Wentang became the founding members of the trilobite palaeontology in China. Generations of NIGPAS scientists have played an important role in trilobite studies in China and made significant contribution to trilobite systematics, biostratigraphy, palaeobiography and ontology, and their recent approaches on Cambrian chronostratigraphy with trilobites as index fossils are regarded as leading this field in the world.

Agnostus hedini
赫氏球接子
寒武纪,新疆西库鲁克塔格

Cambrian; Western Kuruktag, Xinjiang
NIGPAS Collection #7744

Cruziana
二叶石(三叶虫爬痕)
寒武纪,四川荣经

Cambrian; Rongjing, Sichuan

Neseuretus concavus tenellus
凹沟岛头虫弱沟亚种
早奥陶世,陕西汉中

Early Ordovician; Hanzhong, Shaanxi

Coronocephalus dentatus
齿缘王冠虫
志留纪，湖南龙山

Silurian; Longshan, Hunan
NIGPAS Collection #21634

球接子类 / Agnostid

球接子类是一些特殊的小型三叶虫，个体很小，头尾等大，仅有 2~3 个胸节，多数无眼，却在寒武纪地层划分中起到极为重要的作用。球接子类三叶虫全球广泛分布，可能是浮游生物。它们出现并繁盛在寒武纪，奥陶纪数量很少并于奥陶纪晚期灭绝。南京古生物所科学家对江南斜坡带寒武纪三叶虫动物化石群，特别是球接子类三叶虫化石进行了长期细致的研究，使这些三叶虫化石在年代地层学中的全球界线层型剖面和点位（俗称"金钉子"）定义中起到了关键作用，包括寒武系排碧阶（芙蓉统）、古丈阶、江山阶等，这些年代地层单元及其在中国建立的界线层型已列入国际年代地层表。

Agnostid is a peculiar group of small trilobites with similar sized head (cephalon) and tail (pygidium), and only 2 or 3 thoracic segments, mostly eyeless. The agnostics are widely distributed, likely planktonic, flourishing in Cambrian and extinct in Late Ordovician. They are very important in Cambrian stratigraphic division and correlation. NIGPAS scientists have carried out extensive investigations of the deep water trilobite faunas in the Jiangnan Belt, South China, and discovered the great significance of agnostid trilobites in defining the chronostratigraphic boundaries of the Cambrian System, i.e., for GSSPs, for the Cambrian Paibian Stage (Furongian Series), the Guzhangian Stage and the Jiangshanian Stage, adopted in the International Chronostratigraphic Chart.

Evolutionary Treasures

Oktavites spiralis
螺旋奥氏笔石
志留纪,陕西

Silurian; Shaanxi
NIGPAS Collection #MI12

笔石 / Graptolites

笔石是一类已灭绝的无脊椎动物,可能属于半索动物纲,因形似描在岩石上的象形文字而得名;最早出现于寒武纪中期,在奥陶纪和志留纪大量繁盛,于石炭纪早期走向灭绝。笔石以营漂浮群体生活为主,分布十分广泛,演化迅速,其化石是奥陶纪和志留纪地层划分对比的重要标准化石。笔石体最初都由一个圆锥形胎管生出,胎管出芽生出第一个胞管,许多胞管接连生长,排成一条,叫笔石枝;一个笔石体包含一个或多个笔石枝。笔石化石一般在泥页岩中,少数在灰岩、硅质岩中保存。自20世纪20年代,我国已有学者对笔石进行零星记载。早期笔石研究的代表人物有孙云铸、许杰、宋叔和、穆恩之等,他们是中国笔石古生物学的重要奠基者。南京古生物研究所在穆恩之带领下在笔石系统学和生物地层学方面取得重要进展,建立了我国寒武纪到早泥盆世的笔石序列。中国古生物学家通过对笔石、牙形类化石等的综合研究,不仅建立了全球最系统的奥陶纪笔石生物地层序列,也为达瑞威尔阶、赫南特阶、大坪阶等全球界线层型剖面和点位("金钉子")在中国的确立做出了重要贡献。

Didymograptellus eobifidus
始二分小对笔石
早奥陶世，重庆城口

Early Ordovcian; Chengkou, Chongqing
NIGPAS Collection #152349

Graptolites are a group of extinct animals of hemichordates; the name is derived from the Greek for "written rock", as many graptolite fossils resemble hieroglyphs written on the rock. Graptolites originated during middle Cambrian, flourished during Ordovician and Silurian periods, and disappeared in early Carboniferous. Graptolites are considered planktonic and colonial in life style and distributed worldwide, very important in stratigraphy as index fossils for the Ordovician and Silurian. The graptolite colony (rhabdosome) grows from an initial cone-shape individual (sicula), subsequent zooids (each housed in a tubular or cup-like theca) form a branch and many branches make a rhabdosome. Graptolite fossils are commonly preserved in shales and mudstones, sometimes in limestones and cherts. Chinese palaeontologists started graptolite studies in the 1920s and pioneering scholars include Sun Yunzhu, Xu Jie, Song Shuhe, Mu Enzhi, et al. At NIGPAS, Mu Enzhi and his team made important contributions to graptolite systematics and biostratigraphy with the establishment of graptolite biozones from Cambrian to Early Devonian in China. By applying multidisciplinary approaches with graptolites, conodonts and other fossil groups, Chinese palaeontologists not only have established the most comprehensive Ordovician graptolite biozonation, but also contributed to the definition of GSSPs for the Darriwilian Stage, the Hirnatian Stage, and the Dapingian Stage.

Palaeozoic

Evolutionary Treasures

Pseudisograptus manubriatus koi
剑柄假等称笔石长钉状亚种
中奥陶世，江西玉山

Middle Ordovician; Yushan, Jiangxi
NIGPAS Collection #136151

Undulograptus austrodentatus
澳洲齿状波曲笔石
中奥陶世，浙江江山

Middle Ordovician; Jiangshan, Zhejiang
NIGPAS Collection #124891

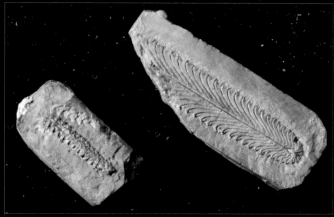

Cardiograptus amplus
长心笔石
早奥陶世，浙江常山

Early Ordovician; Changshan, Zhejiang
NIGPAS Collection #124864

Undulograptus sinodentatus
中国齿状波曲笔石
中奥陶世，浙江江山

Middle Ordovician; Jiangshan, Zhejiang
NIGPAS Collection #124893

Loganograptus logani var. *sinicus*
劳氏劳氏笔石中国变种
早奥陶世，浙江常山

Early Ordovician; Changshan, Zhejiang
NIGPAS Collection #8844

Evolutionary Treasures

腕足动物 / Brachiopoda

腕足动物是生活在海底的有壳无脊椎动物，其两瓣壳大小不同，壳质是钙质或几丁磷灰质。其幼虫期营浮游生活，然后长出肉茎附着于海底、以胶结物或壳刺固着于海底，或自由躺卧海底生活。尤其在古生代，腕足动物极为繁盛并占据海洋动物群的主要地位。腕足动物是具有纤毛环的滤食性动物，貌似双壳（瓣鳃）类软体动物，但两者的亲缘关系甚远。现今腕足动物多数生活于深海和极地水域。腕足类化石非常丰富并广泛分布。

中国腕足动物化石研究的最早奠基人是葛利普，他于1920年刚到中国时就发表了有关二叠纪腕足动物化石的论文。20世纪20年代，赵亚曾对中国石炭纪、二叠纪腕足动物长身贝类做了精深的研究。1931年，葛利普发表了最重要的经典著作《中国泥盆纪腕足类化石》。

Brachiopoda is a large group of marine invertebrates, characterized by a life habit of sessile or resting on substrate, a calacareous or phosphatic shell with two unequal valves, filter-feeding on small organic particles with a lophophore. Especially in the Palaeozoic oceans, brachiopods were extremely rich and dominant. Brachiopod shells are seemingly like those of bivalves, but they are phylogenetically unrelated, the latter belonging to the phylum of Mollusca. Today, brachiopods mostly live in deep water or polar regions, but they were very abundant and widespread in the fossil record.

Brachiopod studies in China began in the 1920s, when Amadeus Grabau published probably the first research paper on fossil brachiopods in China (Permian material from North China); and Zhao Yazeng, a well-known Chinese pioneer in geology and palaeontology, contributed substantially to the studies of Carboniferous and Permian productid brachiopods in China; Grabau published the classic monograph *Devonian Brachiopoda of China* in 1931. These early works have been considered academic frontiers in the world.

Leptodus nobilis
美丽蕉叶贝
二叠纪，贵州毕节

Permian; Bijie, Guizhou
NIGPAS Collection #4641

Evolutionary Treasures

Pentamerus dorsoplanus
背平五房贝
早志留世，湖北宜昌

Early Silurian; Yichang, Hubei
NIGPAS Collection #43953

中国的腕足动物化石研究经过几代人的工作，取得了可观的进展，特别是王钰、金玉玕、戎嘉余等腕足动物研究专家，在该领域取得丰硕成果。例如，有关奥陶纪末期赫南特贝动物群、石燕的起源和早期演化、扭月贝目的系统分类、叶月贝动物群等研究成果在国内外产生广泛影响。王钰、戎嘉余所著《广西南宁—六景间泥盆纪郁江期腕足动物》，金玉玕、戎嘉余等编著的《中国腕足类属志》，金玉玕、戎嘉余等编写的新版《无脊椎古生物学专论》（腕足动物卷）等著作，是我国腕足动物领域的经典文献。

Sinoproductella hemispherica
半球中华小长身贝
晚泥盆世，湖南祁阳

Late Devonian; Qiyang, Hunan
NIGPAS Collection #7448

Generations of Chinese brachiopod experts have substantially advanced the palaeontological subdiscipline. Important figures including Wang Yu, Jin Yugan, Rong Jiayu, et al., have produced tremendous amount of original works of brachiopod research of Cambrian to Permian and Mesozoic. For example, researches on the end-Ordovician Hirnatian Fauna, the origin and evolution of spiriferids, the systematics of Strophomenida, and the Foliomena Fauna have attracted wide academic interest both at home and abroad. Works that may be considered classics or will likely become classics in brachiopology include the monograph *Yukiang (Early Emsian, Devonian) Brachiopods of the Nanning–Liujing District, Central Guangxi, Southern China* by Wang Yu and Rong Jiangyu, *Brachiopod Genera of China* compiled by Jin Yugan, Rong Jiayu, et al., and the new edition of *Treatise on Invertebrate Paleontology (Brachiopoda Part)* by Jin Yugan, Rong Jiayu, et al.

Evolutionary Treasures

Howittia chui
朱氏豪伊特贝
泥盆纪，广西贵县

Devonian; Guixian, Guangxi
NIGPAS Collection #kk30

Stringocephalus sp.
鸮头贝（未定种）
中泥盆世，广西柳江

Middle Devonian; Liujiang, Guangxi

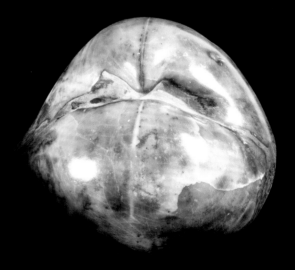

Discoceras sp.
盘角石（未定种）
晚奥陶世，湖北

Late Ordovician; Hubei

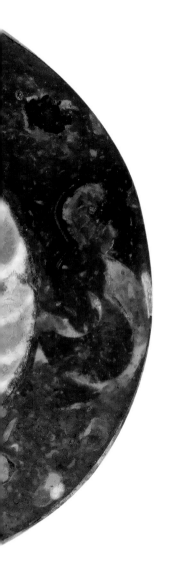

头足动物 / Cephalopoda

头足动物是生活于海洋中的一类软体动物，如鱿鱼和章鱼等都是该家族的成员，最早出现于寒武纪晚期。奥陶纪的海洋是头足动物鹦鹉螺类的天下，这类凶猛的肉食性动物是当时海洋中的霸主。鹦鹉螺类绝大多数营游泳方式生活，其化石分布广泛。奥陶纪鹦鹉螺类化石常见于我国华北板块和华南扬子板块陆表海中。奥陶纪早期的鹦鹉螺类化石多属内角石类，中—晚奥陶世扬子地台区发现的鹦鹉螺类化石有喇叭角石和震旦角石。

1930 年，中国鹦鹉螺类化石研究的早期代表俞建章发表了专著《中国中部奥陶纪头足类化石》。南京古生物所科学家对中国华南、华北和西藏地区的鹦鹉螺类化石开展了广泛的调查研究，为中国鹦鹉螺类的系统分类、生物地层和地理分布以及系统演化等研究做出了重要贡献。

Cephalopoda is a major group of marine Mollusca, including the extant species of squids and octopus. Cephalopods first appeared in late Cambrian and the Ordovician nautiloids dominated the marine biota as excellent swimmers and predators, with extensive fossil record. Nautiloids are very common in epeiric marine Ordovician sediments of the North China Plate and the Yangtze Plate of South China; the Early Ordovician nautiloids mostly belong to the endocerids, while the peculiar coiled *Lituites* and *Sinoceras* are commonly found in the Middle–Late Ordovician strata of the Yangtze Plate.

Pioneering Chinese scholar on nautiloids, Yu Jianzhang published, as early as in 1930, a monograph entitled *The Ordovician Cephalopoda of Central China*. NIGPAS scientists later contributed substantially to nautiloid taxonomy, stratigraphic and geographic distribution in China as well as their evolutionary systematics.

Evolutionary Treasures

Sinoceras chinense
中华震旦角石
晚奥陶世，重庆綦江

Late Ordovician; Qijiang,
Chongqing
NIGPAS Collection #15429

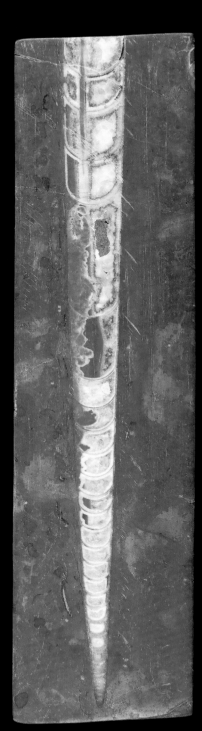

Michelinoceras sp.
米契林角石（未定种）
晚奥陶世，湖北

Late Ordovician; Hubei

Lituites ningkiangensis
宁强喇叭角石
中奥陶世，重庆秀山

Middle Ordovician; Xiushan, Chongqing
NIGPAS Collection #21988

Sinoceras chinense
中华震旦角石
晚奥陶世，湖北宜昌

Late Ordovician; Yichang, Hubei
NIGPAS Collection #2844

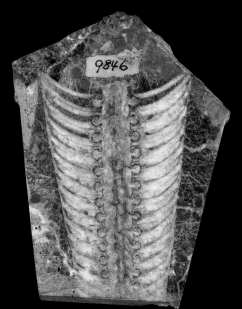

Ordosoceras sphaeriforme
球形鄂尔多斯角石
中奥陶世，内蒙古

Middle Ordovician; Inner Mongolia
NIGPAS Collection #9846

Evolutionary Treasures

腹足动物 / Gastropoda

腹足动物是软体动物门中种类最多的一个纲，常见的有蜗牛、田螺、玉螺等，从寒武纪至第四纪分布广泛，现生种类多达 6 万种。20 世纪 20 年代，葛利普等学者研究中国各门类化石的著作中，已涉及腹足类化石。1929 年，中国早期古生物学者秉志发表了中国腹足类化石的最早著作；此后相继有尹赞勋、许杰、闫敦建等人的专著问世。1949 年之后，余汶等编著了《中国的腹足类化石》（科学出版社，1963），并撰写了一系列论文。

我国的腹足类化石甚为丰富，已公开发表 1300 种之多。西南地区早寒武世地层中发现的原始腹足类化石，对于划分寒武纪早期地层以及探讨腹足类的起源具有重要价值。南京古生物所专家还在晚二叠世陆相腹足类化石以及中生代—新生代腹足类化石研究方面取得了显著进展。

Gastropoda is a major group of Mollusca, commonly known as snails and slugs, both marine and terrestrial inhabitants, with extended fossil record from early Cambrian through Quaternary, including over 60,000 living species. Pioneering palaeontologists in China started recording gastropod fossils as early as in the 1920s; A. Grabau described gastropod fossils in his works of Chinese fossils; Bing Zhi published the earliest monograph on gastropods in 1929, and later scholars Yin Zanxun, Xu Jie, Yan Dunjian et al. published monographic works on gastropods; NIGPAS scientists Yu Wen et al. published numerous original papers and a book entitled *Gastropod Fossils in China* (Science Press, 1963).

Fossil gastropods have been abundantly found in China, with more than 1300 species reported so far, including primitive gastropod fossils from the early Cambrian strata of Southwest China, very important for early Cambrian stratigraphic division and studies of the origin of gastropods. NIGPAS scientists have also conducted extensive investigations of Late Permian terrestrial gastropods and Mesozoic–Cenozoic gastropods.

Strobeus depressus
低圆旋螺
晚石炭世，山西太原

Late Carboniferous; Taiyuan, Shanxi
NIGPAS Collection #4834

Pharkidonotus acutocarinatus
尖棱皱螺
晚石炭世，甘肃临泽

Late Carboniferous; Linze, Gansu
NIGPAS Collection #4771

Naticopsis deformis
变形似玉螺
晚石炭世，山西保德

Late Carboniferous; Baode, Shanxi
NIGPAS Collection #4814

Evolutionary Treasures

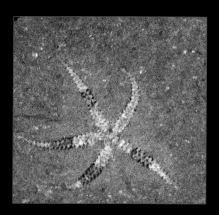
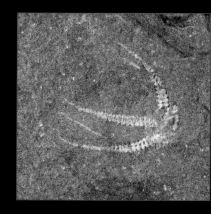

Aganaster? fujianensis
福建阿加尼亚蛇尾(?)
中二叠世，福建龙岩

Middle Permian; Longyan, Fujian
NIGPAS Collection #133175

Caelocrinus stellifer
星状雕刻海百合
早志留世，四川广元

Early Silurian; Guangyuan, Sichuan
NIGPAS Collection #13256

棘皮动物 / Echinodermata

棘皮动物属海生无脊椎动物门，常见类群有海星、海胆、海参和海百合等，包括7000余种现生棘皮动物。

我国棘皮动物化石的研究始于20世纪20年代。1926年，田奇㻛发表了题为《中国北部太原系海百合化石》的专著，这是中国学者所写的第一部也是目前唯一的一部棘皮动物化石专著。之后，孙云铸、穆恩之和杨遵仪等也进行了大量有关棘皮动物化石的研究，内容涵盖广泛，如西藏的海星、海胆和海百合化石，贵州的蛇尾类化石，陕南和湘西的海林檎化石，新疆塔里木西部地区以及西藏班戈等地丰富的海胆化石，以及"关岭生物群"保存完美的海百合化石。近年来，寒武纪棘皮动物化石特别是始海百合类化石在我国西南地区和辽宁大连馒头组中被陆续发现和报道，为研究棘皮动物的起源和早期演化提供了重要证据。

Echinodermata is a group of marine invertebrates, including the extant asteroids (star fish), sea urchins, sea cucumbers and crinoids. Studies of fossil echinoderms began in 1920s in China. Pioneering investigator Tian Qijun published the first monographic work entitled *Crinoids from the Taiyuan Series of North China* in 1926. Later, Sun Yunzhu, Mu Enzhi, Yang Zunyi and others conducted extensive studies of echinoderm fossils, including asteroids, crinoids and sea urchins from Tibet, ophiuroids from Guizhou, cystoids from southern Shaanxi and western Hunan, abundant crinoids from western Tarim Basin (Xinjiang) and northwestern Tibet, and exceptionally preserved crinoids from Guanling Biota (Guizhou). In recent years, the Cambrian echinoderm, especially the eocrinoids, have been discovered in Guizhou and Yunnan in southwestern China and Liaoning in North China, which are especially important for studies of echinoderm's early evolutionary history.

Evolutionary Treasures

Sinocrinus microgranulosus pentalobosus
小多突起中国海百合五板亚种
晚石炭世，河北临城

Late Carboniferous; Lincheng, Hebei
NIGPAS Collection #829

Sinopetalocrinus involutus
包卷中华花瓣海百合
早志留世，贵州石阡

Early Silurian; Shiqian, Guizhou
NIGPAS Collection #73967

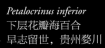

Petalocrinus inferior
下层花瓣海百合
早志留世，贵州婺川

Early Silurian; Wuchuan, Guizhou
NIGPAS Collection #6886

Petalocrinus magnus
大型花瓣海百合
早志留世，贵州石阡

Early Silurian; Shiqian, Guizhou
NIGPAS Collection #73956

珊瑚 / Corals

珊瑚是海生无脊椎动物，属于刺丝胞动物门的一个纲。珊瑚动物通常以群体生长，是重要的热带海洋中的造礁生物，它们能分泌钙质骨骼。珊瑚化石研究是我国古生物研究领域的传统强项。1922 年，葛利普发表了经典著作《中国古生代珊瑚化石》。1926 年，乐森璕发表了中国第一篇珊瑚化石研究论文"奉天直隶石炭纪管状珊瑚虫之新属"。其后，乐森璕、黄汲清、俞建章、计荣森、王鸿祯等进行了大量的研究，建立了中国古生代四射珊瑚的分类体系。1949 年以后，经过俞建章、乐森璕、黄汲清、王鸿祯等的研究工作，在大量属种描述和资料积累的基础上，初步建立起自奥陶纪到二叠纪各纪地层中的珊瑚化石序列。青藏科考及云南红层的研究成果填补了中生代—新生代六射珊瑚化石的空白。

Corals are marine invertebrates, belonging to a class (Anthozoa) of Phylum Cnidaria. They typically live in compact colonies composed of numerous individuals. Corals are important reef builders in tropical oceans; they are able to secrete calcium carbonate to form a hard skeleton.

Studies of coral fossils were traditionally a very strong area in Chinese palaeontology. A. Grabau published the classic work in China in 1922, entitled Palaeozoic Corals of China, followed by Yue Senxun who published the first Chinese research paper on fossil corals "On a New Genus of Syringoporoid Coral from the Carboniferous of Chihli and Fengtien Provinces" in 1926. Later studies by Yue Senxun, Huang Jiqing, Yu Jianzhang, Ji Rongsen, Wang Hongzhen and others laid down the foundation of the taxonomy of Palaeozoic Rugosa corals in China. After the founding of the People's Republic of China, massive accumulation of data and descriptions by teams of palaeontologists, including Yu Jianzhang, Yue Senxun, Huang Jiqing, Wang Hongzhen et al., resulted in the establishment of coral biostratigraphic sequences from Ordovician through Permian. Expeditions of the Qinghai-Tibet Plateau and the investigation of the red strata in Yunnan further filled the gaps in hexacorals of Mesozoic and Cenozoic ages.

Agetolites sp.
阿盖特珊瑚（未定种）
奥陶纪—志留纪，青海

Ordovician–Silurian; Qinghai

Evolutionary Treasures

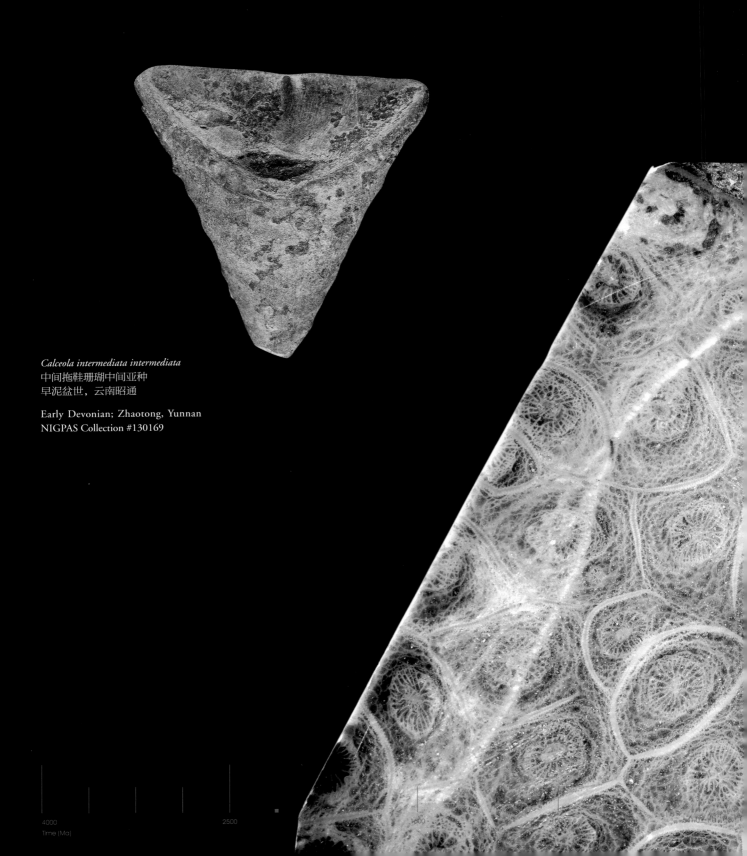

Calceola intermediata intermediata
中间拖鞋珊瑚中间亚种
早泥盆世，云南昭通

Early Devonian; Zhaotong, Yunnan
NIGPAS Collection #130169

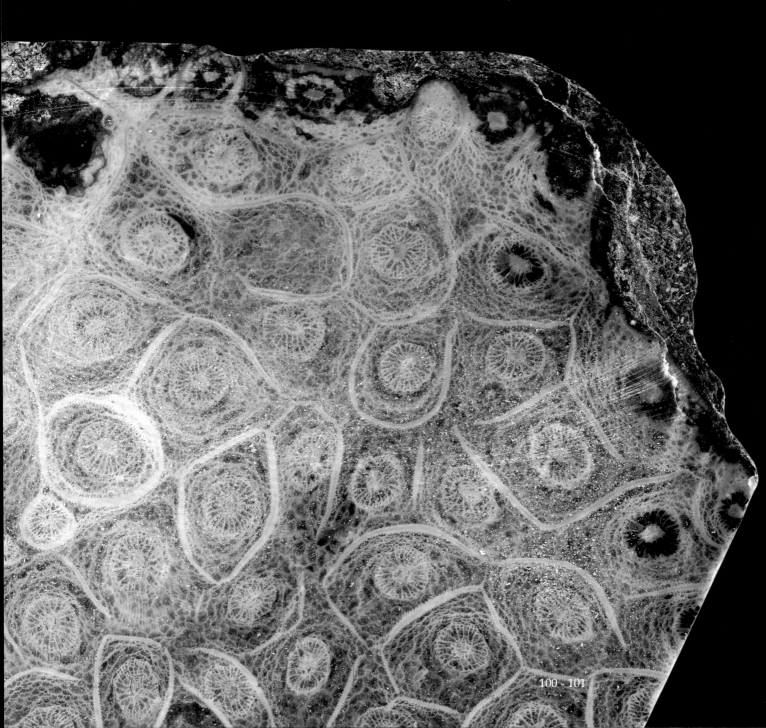

Polythecalis sp.
多壁珊瑚（未定种）
中二叠世，湖南

Middle Permian; Hunan

Evolutionary Treasures

Disphyllum sp.
分珊瑚（未定种）
泥盆纪，湖南

Devonian; Hunan

Palaeosmilia sp.
古剑珊瑚（未定种）
早石炭世，新疆

Early Carboniferous; Xinjiang

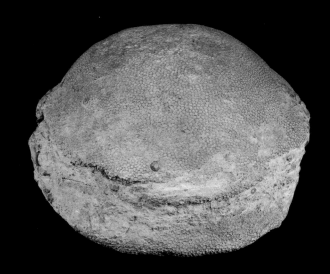

Favosites sp.
蜂巢珊瑚（未定种）
志留纪—泥盆纪，广西

Silurian–Devonian; Guangxi

Evolutionary Treasures

蜓 / Fusulinid

蜓属于原生动物门、有孔虫纲，是一类已灭绝的海洋单细胞微小生物，时代分布局限于古生代晚期的石炭纪和二叠纪；其外形似纺锤（纺织机上的梭子），故而又称为"纺锤虫"。蜓的中文名由李四光创造。1923年，李四光发表的有关蜓类的研究论文是"中国人最早发表的古生物学学术文章"之一。李四光提出的蜓类分类原则被沿用至今。1931年，李四光发表了有关中国石炭—二叠纪蜓类有孔虫的种类和分布的论文，建立了48个蜓类化石带。1934年，陈旭发表了英文著作《中国南部之蜓科》，详细描述了长江下游石炭—二叠纪蜓类14属67种8变种。1949年以后，盛金章等对中国各区域含蜓类化石的地层开展全面研究，他们所发表的《辽宁太子河流域本溪统蜓科》（1958）及《广西、贵州及四川二叠纪的蜓类》（1963），基本确定了中国蜓类的地质、地理分布规律以及蜓类生物地层学框架。

Fusulinid is an extinct group of marine protists, belonging to Class Foraminifera, limited to Carboniferous and Permian marine strata. Li Siguang, who published the first paper on fusulinids in China in 1923, coined its Chinese name. The taxonomic criteria and classification of fusulinids proposed by Li Siguang have been widely adopted. In his paper of fusulinid taxa and distribution in Carboniferous and Permian strata in China published in 1931, Li Siguang established 48 fusulinid biozones. In 1934, Chen Xu published a classic work entitled Fusulinidae of South China with detailed description of 14 genera, 67 species and 8 varieties of fusulinids from the Lower Yangtze region. After the founding of the People's Republic of China, Sheng Jinzhang and others conducted extensive investigation of fusulinid faunas in various regions of China, and published the widely influencing monographs including Fusulinids from the Penchi Series of the Taitzeho Valley, Liaoning (1958), and Permian Fusulinids of Kwangsi, Kueichow and Szechuan (1963), setting a sound foundation for the geological and geographical distribution of fusulinids in China and their biostratigraphic framework.

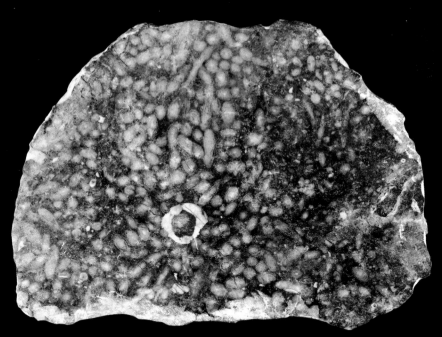

Schwagerina sp.
希瓦格蜓（未定种）
晚石炭世，新疆柯坪

Late Carboniferous; Kalpin, Xinjiang

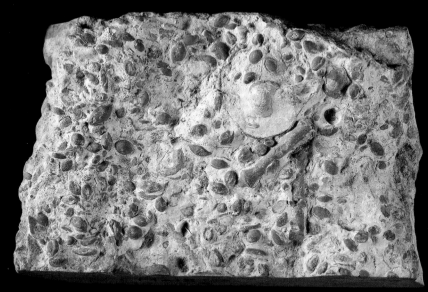

Pseudofusulina sp.
假纺锤蜓（未定种）
晚石炭世，新疆

Late Carboniferous; Xinjiang

Fenestella sp.
窗格苔藓虫（未定种）
石炭纪—二叠纪，新疆
Permian—Carboniferous; Xinjiang
NIGPAS Collection #MI94

苔藓虫 / Bryozoans

苔藓虫是一类水生滤食性生物，除一个属为单体生活外，其余均为群体生活；典型个体大小约 0.5 mm；仅少数属种营淡水生活，绝大多数生活于海洋之中；能分泌钙质或几丁质外骨骼。苔藓虫群体呈扇状、枝状或层状附着于基岩、岩屑、海藻等物体之上；多数群体大小为 1~10 cm，有些群体可达 1 m。苔藓虫骨骼化石广泛分布于奥陶纪以来世界各地不同时代的地层中。

20 世纪 50 年代初，南京古生物所杨敬之等开展了中国苔藓虫化石研究，对四川、广西、湖北、湖南、浙江、陕西、吉林及西藏等地区奥陶纪至三叠纪苔藓虫化石进行了广泛的调查研究，发表了一系列地层学和系统古生物学论文，编著出版了《中国的苔藓虫》《苔藓虫化石》（古生物学小丛书）及《珠穆朗玛峰地区的苔藓虫化石》。近年来，南京古生物所科学家在湖北宜昌黄花场地区的分乡组中，发现了世界上最古老的特马豆克期苔藓虫化石，引起国际同行的关注。

Bryozoan is a group of aquatic invertebrate animals typically 0.5 mm long. They are filter feeders obtaining food particles from the water with a retractable lophophore. Only one family of species are fresh water dwellers, all the others are marine. Bryozoans can secrete exoskeletons of chitins or calcium carbonate; only one genus is solitary while all other bryozoans are colonials, forming fans, bushes or sheets attached to a hard substrate such as surfaces of a rock, algae or other objects; colonies are normally 1-10 cm in size, some may reach 1 m. Bryozoan fossils are distributed worldwide from the Ordovician and younger strata.

Palaeontological studies of Bryozoa began in 1950s by Yang Jingzhi and his team members at NIGPAS. They conducted extensive investigation of Ordovician through Triassic strata nationwide, including Sichuan, Guangxi, Hubei, Hunan, Zhejiang, Shaanxi, Jilin and Tibet, with the publications of numerous books and research papers, such as *Bryozoans in China*, *Bryozoan Fossils* (for non-professionals) and Bryozoan Fossils for the Mt. Qomolangma Region. Recently, NIGPAS scientists discovered the earliest fossil bryozoans of Ordovician Tremadocian Age in Huanghuachang, Yichang, Hubei, which attracted international attentions.

Evolutionary Treasures

Cavuspganthus convexus
凸形凹颚刺
石炭纪维宪期，贵州

Carboniferous Visean Age; Guizhou

Gnathodus bilineatus bilineatus
双线颚齿刺双线亚种
石炭纪维宪期，贵州

Carboniferous Visean Age; Guizhou

牙形类 / Conodonts

牙形类是一类已灭绝的海洋原始脊椎动物，形似泥鳅，个体大小为 1~40 cm。化石记录中，很少发现完整的动物个体，而大量保存的是牙形类的磷灰质器官，称为"牙形石"或"牙形刺"。牙形类化石广泛存在于寒武纪到三叠纪海相地层中。牙形类大小约为 1 mm，最大 7 mm，形态多样，具有十分重要的地层学意义。

我国牙形类化石研究始于 20 世纪 50 年代后期。1960 年，南京古生物所金玉玕发表了中国第一篇牙形类论文。1973 年始，全国科研机构、高校、石油和地质系统的微体古生物工作者对全国各区域含牙形刺地层进行了系统采集和研究，为该领域的发展打下了良好的基础。至 20 世纪 90 年代初，我国已建立了从寒武纪到三叠纪的 155 个牙形刺生物带。牙形刺生物地层学在区域地质调查、油气勘探开发、全球地层界线层型和点位的研究中发挥了十分重要的作用。

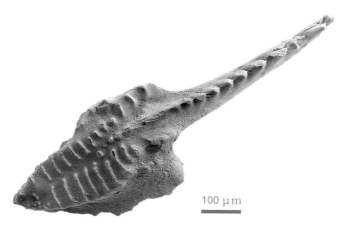

Idiognathodus simulator
仿效异颚刺
石炭纪格舍尔期,贵州

Carboniferous Gzhelian Age; Guizhou

Neognathodus symmetricus
对称新颚齿刺
石炭纪巴什基尔期,贵州

Carboniferous Bashkirian Age; Guizhou

Conodonts are extinct, marine primitive vertebrates, eel-like, 1–40 cm in size. Whole conodont animals are rarely found, but their phosphate skeletal elements (organs) widely and commonly occur in Cambrian through Triassic marine strata. Conodont skeletal elements are generally about 1 mm (max. 7 mm) in size, of variable morphology and extremely useful stratigraphically.

The conodont studies began in the late 1950s in China. NIGPAS palaeontologist Jin Yugan published the first research paper on conodonts in 1960 and the research area experienced very rapid growth. Beginning in 1973, Chinese micropalaeontologists from research institutions, universities, geological surveys and petroleum corporations launched extensive investigations of conodonts-bearing strata in various regions of China, laying down a solid foundation for the discipline's further development. By early 1990s, some 155 conodont biozones were established for the Cambrian to Triassic stratigraphy. It is generally recognized that conodont biostratigraphy in combination with other interdisciplinary approaches has played an exceptional role in regional geological survey, exploration and development of oil and gas, and global stratigraphic timescale.

Evolutionary Treasures

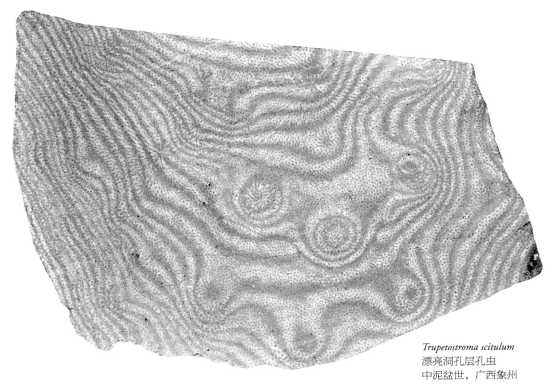

Trupetostroma scitulum
漂亮洞孔层孔虫
中泥盆世，广西象州

Middle Devonian; Xiangzhou, Guangxi
NIGPAS Collection #32927

层孔虫 / Stromatoporoidea

层孔虫是海绵动物门的一个纲，为底栖固着的海洋生物。它是一类重要的造礁生物，出现于早奥陶世，于晚泥盆世开始衰退，白垩纪时灭绝。层孔虫主要生活在热带至亚热带的清浅、正常盐度、水动力条件较强、光照较好的海水环境中，因此，层孔虫化石可作为判断沉积环境的标志化石。

中国层孔虫化石的早期研究包括计荣森（1940）发表的关于西南地区志留纪、泥盆纪层孔虫化石，以及俞建章（1947）关于桂林附近泥盆纪层孔虫化石的分类研究等。1949年以后，南京古生物所杨敬之领导中国研究者进行了系统的层孔虫研究。1962年，南京古生物所杨敬之等编著出版了《中国的层孔虫》一书，成立了中国的层孔虫—苔藓虫专业组，并建立了我国层孔虫化石的新分类系统，其时代从奥陶纪至白垩纪，研究区域遍及我国主要省区。

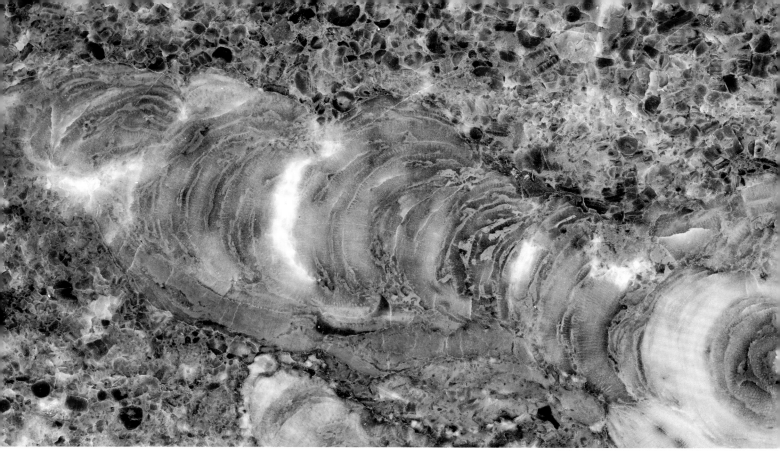

Stromatopora sp.
层孔虫（未定种）
泥盆纪，广西

Devonian; Guangxi

Stromatoporoidea is a group (Class) of Porifera (sponges), commonly occurring as reef-building fossils in Ordovician to Devonian strata; their diversity and abundance rapidly declined after Late Devonian and the group became extinct during Cretaceous Period. Stromatoporoids are sessile marine metazoans and live primarily in tropical to subtropical shallow sea within the photic zone of normal salinity and water currents, thus being a useful indicator for depositional environment.

Early studies of stromatoporoids in China include Ji Rongsen (1940) on Silurian and Devonian stromatoporoid fossils from southwest China and Yu Jianzhang (1947) on Devonian stromatoporoids from Guilin area of Guangxi. After 1949, NIGPAS palaeontologist Yang Jingzhi organized extensive investigation of stromatoporoids-bearing strata in China. Yang Jingzhi et al. published the monograph Stromatoporoids in China; a professional group of Stromatoporoidea–Bryzoa was formed in China, the new taxonomic systematics of stromatoporoids in China was established, and the researches covered Ordovician through Cretaceous strata in major provincial regions in China.

晚古生代动植物登陆景观复原图
（南京古生物博物馆）

Imaginary scene of Late Palaeozoic terrestrialization
(Nanjing Museum of Palaeontology)

维管植物的起源与早期演化

维管植物的起源和早期演化是地球陆地生态系统形成中的最为重要的事件。化石记录显示，陆生维管植物的多样化始于志留纪晚期至泥盆纪早期，维管植物的根、叶等器官大量出现于泥盆纪中期，大型树木和森林以及原始的种子蕨类植物出现于泥盆纪晚期。

我国的早期维管植物化石研究始于 20 世纪 30 年代。瑞典学者赫勒（T.G. Halle）于 1936 年发表了云南古生代植物化石的研究成果。中国古植物学开创者、南京古生物所的斯行健（1936，1937）发表了两篇关于湖南早期陆生植物化石的研究成果，并于 1941 年发表了关于云南昭通早期陆生植物化石的研究成果。20 世纪 60 和 70 年代，研究者在很多地区发现了大量早期陆生维管植物化石。近 30 年来，我国科学家在早期陆生维管植物化石研究领域取得了新进展，特别是新疆志留纪末期和中泥盆世植物群、云南早泥盆世坡松冲和徐家冲植物群、云南中泥盆世晚期西冲植物群以及长江中下游晚泥盆世晚期五通植物群的发现与研究。

The Origin and Early Evolution of Vascular Plants

The origin and early evolution of vascular plants is probably the most important evolutionary event in the formation of the terrestrial ecosystem on the Earth. The fossil record indicates that this diversification of vascular plants began during Late Silurian to Early Devonian times. By Middle Devonian time, features found in today's plants such as roots and leaves are widely found in the sediments and Late Devonian plants include forest-forming large woody trees and primitive seed ferns.

Studies of early vascular plants in China began in the 1930s, when Swedish scholar T.G. Halle (1936) reported findings of Palaeozoic plant fossils from Yunnan and pioneering Chinese palaeobotanist Si Xingjian (1936, 1937) of NIGPAS published studies of early terrestrial plants from Hunan. Si Xingjian (1941) again published his study of such fossils from Zhaotong, Yunnan. During the 1960s and 1970s, early vascular plant fossils were discovered nationwide. During recent 30 years, remarkable advances in this research area has been made in China; particularly worth mentioning are the discoveries of early land vascular plants from latest Silurian and Middle Devonian strata of Xinjiang, the Early Devonian Posongchong and Xujiachong floras of Yunnan, and the Late Devonian Wutong flora of the Lower Yangtze region.

Evolutionary Treasures

Archaeopteris roemeriana
罗氏古羊齿
晚泥盆世，广东新会

Late Devonian; Xinhui, Guangdong
NIGPAS Collection #PB11728

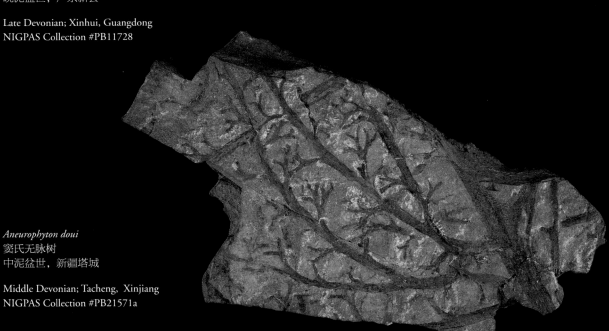

Aneurophyton doui
窦氏无脉树
中泥盆世，新疆塔城

Middle Devonian; Tacheng, Xinjiang
NIGPAS Collection #PB21571a

Wuxia bistrobilata
双穗无锡蕨
晚泥盆世，江苏无锡

Late Devonian; Wuxi, Jiangsu
NIGPAS Collection #PB18870

Neuropteris gigantea
大脉羊齿
晚石炭世，江西乐平

Late Carboniferous; Leping, Jiangxi
NIGPAS Collection #PB2621

Evolutionary Treasures

Pecopteris mui
穆氏栉羊齿
早石炭世，青海

Early Carboniferous; Qinghai
NIGPAS Collection #PB2712

成煤时代的植物 / Plants of Coal-Forming Period

石炭纪时期，地球上气候温暖潮湿，陆生维管植物十分繁盛，高大的乔木在气候潮湿的陆地上形成规模巨大的原始森林，为石炭纪煤层的形成提供了雄厚的物质基础。鳞木、拟鳞木、封印木、古芦木、芦木、轮叶、科达、脉羊齿、栉羊齿等持续发展，种类繁多。晚石炭世，随着全球气候格局的变化，出现了明显的植物地理分区，安加拉植物群、冈瓦纳植物群、欧美植物群和华夏植物群各据一方。二叠纪中后期，气候慢慢变干，原先适应温暖潮湿环境的植物开始消退，裸子植物逐渐占据了主导地位，松柏和苏铁类十分繁盛。

During Carboniferous Period, due to the prevailing warm and humid climate on the Earth, land vascular plants were prosperous, with tall trees forming vast forests that provided the sufficient material base for the formation of abundant coals. The floras were highly diverse with *Lepidodendron, Lepidodendropsis, Sigillaria, Archaeocalamites, Calamites, Annularia, Cordaites, Neuropteris, Pecopteris*, and widely developing. Because of the Late Carboniferous global climate change, distinct phytogeographic provinces were formed, including the Angaran Flora, the Gondwanan Flora, the Euramerican Flora and Cathaysian Flora. During Late Permian, due to global climate change with increasing aridity, gymnosperms gradually replaced earlier floras of warm and wet weather, with flourishing conifers and cycads dominant.

Evolutionary Treasures

Lepidodendron aolungpylukense
鱼鳞木
晚石炭世，青海德令哈

Late Carboniferous; Delingha, Qinghai
NIGPAS Collection #PB2741

Sinocycadoxylon liianum
李氏中国苏铁木
中侏罗世,辽宁

Middle Jurassic; Liaoning
NIGPAS Collection #PB21393

Lepidodendron ochluisfetis
猫眼鳞木
晚石炭世，山西武乡
Late Carboniferous; Wuxiang/Shanxi
NIGPAS Collection#PB3054

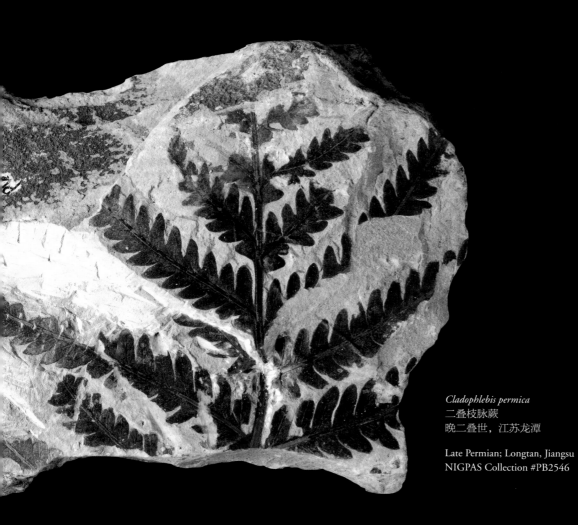

Cladophlebis permica
二叠枝脉蕨
晚二叠世，江苏龙潭

Late Permian; Longtan, Jiangsu
NIGPAS Collection #PB2546

Crossotheca cathaysiana
华夏穗囊蕨
早二叠世，山西山阴

Early Permian; Shanyin, Shanxi
NIGPAS Collection #PB11216

Evolutionary Treasures

Annularia gracilescens
纤细轮叶
早—中二叠世，山西河曲

Early–Middle Permian; Hequ, Shanxi
NIGPAS Collection #PB3888

Evolutionary Treasures

华夏植物群 / Cathaysian Flora

石炭纪和二叠纪，地球上形成了 4 个重要的植物区系：欧美植物群、华夏植物群、安加拉植物群和冈瓦纳植物群。华夏植物群主要分布在中国的华北、华南，日本、朝鲜半岛以及东南亚等地区。华夏植物群的代表植物是大羽羊齿，同时繁盛的还有华夏羊齿、织羊齿、瓣轮叶等。南京古生物所斯行健、李星学带领中国古生物学者，从 20 世纪 50 年代开始对石炭纪和二叠纪植物群进行了深入的研究，提出了"前华夏植物群"的概念，对探讨华夏植物群起源与演化具有重要意义。几代古植物学工作者经过研究，证明了华夏植物群自晚石炭世初期形成并逐渐繁盛，至二叠纪末消亡，历经 6000 多万年演化过程；确定了华夏植物群北与安加拉植物群、西南与冈瓦纳植物群之间的分界。

During Carboniferous and Permian periods, there formed 4 major phytogeographic provinces, namely, the Euramerican Flora, the Cathaysian Flora, the Angaran Flora and the Gondwanan Flora. The distribution area for the Cathaysian Flora included North China, South China, Japan, Korean Peninsula and parts of Southeast Asia, all characterized by the presence of *Gigantopteris*; other common constituent plants of the Cathaysian Flora include *Cathaysiopteris, Emplectopteris and Lobatannularia*. Since the 1950s, NIGPAS scientists Si Xingjian and Li Xingxue along with their fellow palaeobotanists conducted extensive investigation of the Carboniferous and Permian floras and proposed the concept of "Pre-Cathaysian Flora", a significant contribution to understanding the origin and evolution of the Cathaysian Flora. In recent years, NIGPAS palaeobotanists contributed significantly to studies of the Cathaysian Flora. After continued studies by generations of Chinese palaeobotanists, it is now understood that the Cathaysian Flora formed during the beginning of Late Carboniferous Period, became widely flourishing afterwards, and went to extinction by the end of Permian Period, spanning over 60 million years' evolution; it is also clear about the boundaries between the Cathaysian Flora and the Angaran Flora to the north or the Gondwanan Flora to the southwest.

Paratingia wudensis
乌达拟齿叶
二叠纪，内蒙古乌达

Permian; Wuda, Inner Mongolia
NIGPAS Collection #PB20776

Evolutionary Treasures

Sphenophyllum thonii
畸楔叶
二叠纪，山西

Permian; Shanxi
NIGPAS Collection #PB3041

Tingia carbonica
华夏齿叶
中二叠世，山西保德

Middle Permian; Baode, Shanxi
NIGPAS Collection #PB21242

Evolutionary Treasures

Pecopteris lativenosa
厚脉栉羊齿
二叠纪，内蒙古乌达

Permian; Wuda, Inner Mongolia
NIGPAS Collection #PB21059

植物"庞贝城" / The Vegetational "Pompeii"

近年来，南京古生物所研究人员发现，在内蒙古乌达煤田，存在距今约3亿年的被火山喷发所埋藏的沼泽森林。通过大量化石发掘和细致的埋藏学分析，科学家成功再现了远古森林的原始景观和植物组成。研究表明，该森林的群落结构保存完美，主要由六大植物类群组成：石松类、有节类、瓢叶类、蕨类、原始松柏类和苏铁类。高层植被由原始松柏类科达和石松类封印木构成，中层植被的蕨类植物构成了森林的主体，而底层植被包括有节类植物楔叶和星叶等。由于当时植物群所处环境有着地球处于冰室—温室过渡的气候背景，因此，这项研究对探讨现代植被随气候变化的趋势具有重要的参考价值。

In recent years, NIGPAS palaeobotanists discovered from the Inner Monglia Wuda coal mines, a magnificent swamp forest of some 300 million years in age, buried by volcanic ashes. Through large scale excavation and *in situ* taphonomic studies, scientists were able to reconstruct the actual landscapes and palaeobotanical features of the forest, revealing an exceptionally preserved forest community structure and composition. Six major constituent plant groups were found to be lycopsids, articulates, noeggerathiopsids, ferns, early coniferophytes, and cycads; the upper tier of the forest consists of cordaitopsids and sagitarians, the middle tier of tree ferns, and the lower tier of sphenopsids and asterophyllites. The Wuda Flora represents a time interval of global climatic transition from ice-house to green-house conditions, which is considered useful for understanding the current floral change in response to the putative global warming.

Sigillaria cf. *ichthyolepis*
鱼鳞封印木（比较种）
二叠纪，内蒙古乌达

Permian; Wuda, Inner Mongolia
NIGPAS Collection #PB21429

中生代

Mesozoic:

The Lost World and the Forerunners of the Modern Ecosphere

Phanerozoic >

Evolutionary Treasures

大约 2.52 亿年前，地球历史进入一个承前启后的新时代——中生代。中生代包括三个纪：三叠纪（2.52 亿~2.01 亿年前）、侏罗纪（2.01 亿~1.45 亿年前）、白垩纪（1.45 亿~0.66 亿年前），共持续了大约 1.86 亿年。中生代被称为"裸子植物时代"、"恐龙时代"、"菊石时代"，恐龙和裸子植物是陆地上的主角，而菊石动物在海洋中称雄；海洋中还出现了大量海生爬行类。哺乳动物起源于中生代，它们几乎与恐龙同行，但却一直生活在恐龙的阴影中。陆地上，蕨类植物和被子植物则处在裸子植物的阴影之中——虽然蕨类植物在石炭纪和二叠纪是陆地植被中的主角，但到了中生代，由于气候变化等原因，裸子植物占据主导地位。中生代哺乳动物和被子植物的演化起源是生命演化史上的重大事件，近年来中国古生物学工作者在该领域的研究中取得了重要进展。由于陆地的不断扩大，中国区域内淡水无脊椎动物蓬勃发展，双壳类、腹足类、介形类、叶肢介等无脊椎动物十分繁盛，其化石成为陆相地层划分与对比以及地质和化石能源勘探研究中的重要依据。

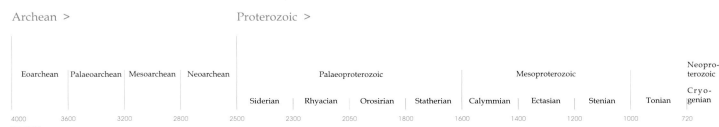

About 252 Ma, the Earth entered a brand new era of Mesozoic, which is divided into three periods, namely, the Triassic (ca. 252–201 Ma), the Jurassic (201–145 Ma) and the Cretaceous (145–66 Ma), spanning a time interval of some 186 million years. The Mesozoic is often referred to as the "Age of Dinosaurs", the "Age of Gymnosperms", or the "Age of Ammonites", because dinosaurs and gymnosperms (such as pines and cycads) were dominating the terrestrial biotas, while ammonites were super abundant in the marine realms, along with thriving and diversifying marine reptiles. Similarly, ferns and angiosperms lived in the shadows of gymnosperms. Early mammals evolved during the Mesozoic in the shadows of prosperous and dominating dinosaur faunas. While ferns prospered during Late Palaeozoic, their predominant role was replaced by gymnosperms and the climatic conditions also changed dramatically. The origin and evolution of angiosperms and mammals during the Mesozoic are two major evolutionary events in the life history; for both events, significant advances have been made by Chinese palaeontologists in the past decades. Because the expansion of terrestrial sedimentary regimes in Chinese territories, abundant freshwater invertebrate faunas occur, including bivalves, gastropods, ostracodes and conchostracans, which have played very important roles in stratigraphic division and correlation of terrestrial sediments, and in geological exploration of natural resources, especially the fossil fuels (oil and gas).

Phanerozoic >

Evolutionary Treasures

Guizhouella analilepida
臀鳞贵州真颌鱼
晚三叠世，贵州兴义

Late Triassic; Xingyi, Guizhou
NIGPAS Collection #136041

"关岭生物群" / "Guanling Biota"

"关岭生物群"（距今约 2.2 亿年）产于贵州关岭新铺乡附近的上三叠统小凹组灰黑色灰泥岩中，以大量保存精美的海生爬行类（鱼龙、海龙、鳍龙、盾齿龙等）和棘皮动物海百合化石为特征，此外还有丰富的鱼类、菊石、双壳类、腹足类、牙形类和植物化石，是一个珍稀的特异埋藏生物化石群。穆恩之于 1949 年首次发表了"关岭生物群"中海百合化石的研究成果。近年来，古生物学家在黔西南—滇东北相邻区域发现了大量与关岭生物群相关联的中—晚三叠世海洋生物群，包括"兴义生物群"（贵州兴义，中—晚三叠世，竹杆坡组，产出大量贵州龙化石）、"盘县生物群"（贵州盘县，中三叠世，关岭组上段，产出保存精美的鱼龙类、幻龙类、原龙类等海生爬行类群化石）、"罗平生物群"（云南罗平，中三叠世，关岭组，产出丰富的海生鱼类化石，伴有海生爬行类、棘皮类、甲壳类、软体动物和裸子植物化石）。上述黔西南—滇东北地区发育的中—晚三叠世特异埋藏化石宝库，代表了扬子板块边缘海域中的较深水沉积盆地，形成于印支构造期。三叠纪晚期，东南方向的南盘江盆地闭合，形成相对封闭的沉积盆地，而沉积盆地底部环境缺氧，有利于化石的大量保存。

Qianichyosaurus zhoui
周氏黔鱼龙
晚三叠世，贵州关岭

Late Triassic; Guanling, Guizhou

The "Guanling Biota" is discovered near the township of Xinpu, Guanling, Guizhou, in Late Triassic dark mainly limestone or mudstone of the Xiao'ao Formation, dated about 220 Ma. The Guanling Biota is well-known for its superbly preserved marine vertebrates (including ichthyosaurs, thalattosaurs, sauropterygians, placodonts, etc.) and crinoid fossils, along with abundant marine fishes, ammonites, bivalves, gastropods, conodonts and terrestrial plants. NIGPAS scientist Mu Enzhi (1949) published the first scientific paper on the Guanling Biota with descriptions of crinoid fossils collected therein. In recent years, Chinese palaeontologists have devoted great effort with magnificent new discoveries of Middle and Late Triassic marine Konservat-Lagerstätten, in addition to the Guangling Biota, from the region of southwestern Guizhou and northeastern Yunnan provinces, including the Xingyi Biota found from the Middle–Late Triassic Zhuganpo Formation in Xingyi, Guizhou, featuring the publically famous Keichousaurus, the Panxian Biota from Panxian, Guizhou, found in the upper part of the Middle Triassic Guanling Formation, characterized by exceptionally preserved ichthyosaurs, nothosaurs and protorosaurs, and the Luoping Biota from Luoping, Yunnan, and also found in the Middle Triassic Guanling Formation, featuring very abundant marine fishes, along with marine reptiles, echinoderms, crustaceans, molluscs, and plant fossils.

Guizhouichthyosaurus tangae
邓氏贵州鱼龙
三叠纪，贵州关岭

Triassic; Guanling, Guizhou

Time (Ma)
4000　　　　　　2500　　　　　　1600　　　　1000　　　541.0 ± 1.0

邓氏贵州鱼龙复原图
Cartoon of *Guizhouichthyosaurus tangae*

Evolutionary Treasures

菊石 / Ammonites

菊石是一类已经灭绝的头足纲海洋软体动物，与鹦鹉螺、章鱼、乌贼等是近亲。菊石出现于泥盆纪，白垩纪末与恐龙同时从地球上消失。菊石化石分布广泛，特征鲜明，演化速率很快，地层学家经常用其作为划分和对比地层的标准化石。化石爱好者将菊石作为观赏化石。

我国早期研究菊石化石的学者有尹赞勋、田奇瑰、赵金科和许德佑等。南京古生物所前所长赵金科带领研究组成员在中国对菊石化石开展了广泛的探索和研究，建立了中国菊石化石序列。赵金科等发表的《华南晚二叠世头足类》（1978），系统描述了华南 80 余产地 155 种菊石和 16 种鹦鹉螺化石，提出了"华夏动物群"的概念。2009 年，南京古生物所科学家参与了新版 Treatise on Invertebrate Paleontology 头足类卷的编著，书中描述了全球石炭纪和二叠纪菊石 426 属，收录了中国南方形态奇异的土著属，为全球菊石化石研究做出了重要贡献。

Ammonoites are an extinct group of marine molluscs belonging to Class Cephalopoda, related to today's *Nautilus*, and more closely related to squids, octopus and cuttlefish. Ammonites first appeared during Devonian times, most abundant in Mesozoic oceans, and terminated, together with dinosaurs, in the end Cretaceous mass extinction event. Ammonites are excellent index fossils, characterized by their abundance and worldwide distribution in marine sediments, and thus are very important in stratigraphic correlation. They are also commonly favored by fossil fans.

Pioneering ammonite researchers in China include Yin Zanxun, Tian Qijun, Zhao Jinke and Xu Deyou. NIGPAS former director Zhao Jinke and his team conducted extensive investigation in China and formally established an ammonite stratigraphic sequence of China. In his work entitled Late Permian Cephalopoda of South China (1978), Zhao systematically reported over 170 ammonite and nautiloid species from some 80 localities and proposed the concept of Cathaysian Fauna. Recently, NIGPAS scientist coauthored the classic *Treatise on Invertebrate Palaeontology* on Carboniferous and Permian Ammonoidea (2009) that included endemic ammonites from South China.

Strigogoniatites liuchowensis
柳州尖棱腹菊石
中二叠世,广西柳江

Middle Permian; Liujiang, Guangxi
NIGPAS Collection #7468

Evolutionary Treasures

Propopanoceras kueichowense
贵州前饼菊石
早二叠世，贵州郎岱

Early Permian; Langdai, Guizhou
NIGPAS Collection #22029

Branneroceras yohi
乐氏布朗氏菊石
晚石炭世，贵州水城

Late Carboniferous; Shuicheng, Guizhou
NIGPAS Collection #5825

Indojuvavites angulatus
角腹印度侏瓦菊石
晚三叠世，西藏聂拉木

Late Triassic; Nyalam, South Tibet
NIGPAS Collection #31462

146 - 147

4000
Time (Ma)
2500
1600
1000
541.0±1.0

Evolutionary Treasures

海百合类 / Crinoids

海百合类是棘皮动物门的一个纲，海生，包括柄海百合和海羊齿两类；前者成体固着生活（幼体浮游），后者可以在海底移动。海百合的口朝上，周边是一系列羽状腕，用于捕食。海百合化石在古生界非常丰富，许多灰岩中含有大量海百合茎或碎片。海百合在二叠纪末大灭绝事件中受到重创，并在三叠纪发生辐射演化。我国云贵地区中—晚三叠世海相地层中产出大量完整的海百合化石。现生柄海百合生活在浅海至 9000 m 深海，数量较少。

Crinoids are marine animals belonging to Phylum Echinodermata Class Crinoidea, living in shallow to deep waters (up to 9000 m deep). Crinoids are characterized by a mouth facing up, surrounded by feather-like feeding arms, including two major groups: stalked sea lilies and unstalked comatulids; the former is attached to the sea bottom when mature (juveniles are planktonic) while the latter group can move around. They are abundant in the Palaeozoic strata, with many limestones filled with crinoid stems and/or fragments. They suffered a great loss at the end-Permian mass extinction event, and made a second radiation during the Triassic. In the Middle and Late Triassic marine fossil Laggerstätten in Guizhou–Yunnan region, abundant exceptionally preserved crinoid fossils are found.

Traumatocrinus hsui
许氏创孔海百合
中三叠世，贵州关岭

Middle Triassic; Guanling, Guizhou
NIGPAS Collection #6889

Evolutionary Treasures

Balanocidaris dixoni
分叉坚果形头帕海胆
早白垩世，西藏班戈

Early Cretaceous; Baingoin, Tibet
NIGPAS Collection #71440

Epiaster venustus
雅致顶星海胆
晚白垩世，西藏亚东

Late Cretaceous; Yadong, Tibet
NIGPAS Collection #71541

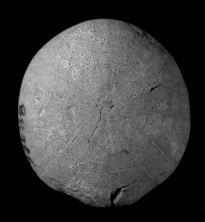

Pygaulus cyclotus
圆尾臀海胆
晚白垩世，新疆乌恰

Late Cretaceous; Wuqia, Xinjiang
NIGPAS Collection #88356

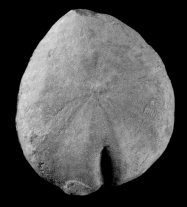

Cardiaster xizangensis
西藏心星海胆
晚白垩世，西藏亚东

Late Cretaceous; Yadong, Tibet
NIGPAS Collection #71491

Evolutionary Treasures

Arguniella lingyuanensis
凌源额尔古纳蚌
早白垩世，辽宁北票
Early Cretaceous; Beipiao, Liaoning

双壳类 / Bivalvia

双壳类又称"瓣鳃类"，是软体动物门的一个纲，生活于海洋或淡水，包括蚌、蛤蜊、扇贝、牡蛎、贻贝等常见贝类；它们多数生活在沉积物表面附近，少数营游泳生活（如扇贝类），是海洋中最丰富的一类动物。双壳类具有对称的钙质双瓣壳（故得名），壳的顶部绞合在一起；它们有一对很大的鳃，发育特殊的梳状鳃瓣（故又名"瓣鳃类"）用于滤食和呼吸。双壳类起源于寒武纪早期，是生命演化史上最为成功的一类动物；二叠纪大灭绝事件以后，双壳类发生了大规模辐射，替代古生代占主导地位的腕足动物，成为滤食动物的主流。

我国双壳类化石研究始于20世纪20年代，先驱研究者包括葛利普、赵亚曾、许德佑等。南京古生物所顾知微领导了中国的瓣鳃类学科方向，在地质、油气勘探、海洋地质等领域开展了广泛探索，建立了一支专业队伍，使得研究工作取得全面进展。他们的重要发现在《中国的瓣鳃类化石》（1976）一书中发表。南京古生物所科学家参与了新版 *Treatise on Invertebrate Paleontology*（《无脊椎动物古生物学》）双壳纲卷的修订工作，并将1927—2007年发表的中国双壳类属和属级以上分类单元汇总发表。

Halobia jomdaensis
江达海燕蛤
晚三叠世，西藏江达

Late Triassic; Gyamda, Tibet
NIGPAS Collection #50852

Bivalvia is also known as "Lamellibranchia" due to their possession of a large pair of gills with specialized structure "ctenidia" for filter-feeding and breathing. They belong to a class of Mollusca, including clams, scallops, oysters, mussels, etc, living in both marine and fresh waters, being most diversified marine animals. Bivalves are so named for their pair of calcareous shells, which are mirror-symmetrically hinged at the upper side with teeth and ligaments. They mostly live on substrate or in sediments, some (such as scallops) can swim. Bivalve fossils date back to early Cambrian, very abundant throughout the geological history up to today; they are thus considered most successful animals in evolution. During Mesozoic, bivalves underwent a great radiation in diversity, becoming the dominant filter-feeding animals, replacing the Palaeozoic dominating brachiopods.

Bivalve studies in China began in the 1920s, with pioneering scholars A. Grabau, Zhao Yazeng, Xu Deyou et al. From 1950s, NIGPAS palaeontologist Gu Zhiwei was key in developing this research discipline, by organizing a strong research team and ventured extensive investigation in geological and petrological exploarations and in marine geology throughout Chinese territories, with remarkable achievements. Most bivalve discoveries were summarized in the 1976 volume of Lamellibranchia Fossils in China. NIGPAS scientists participated in the revision of *Treatise on Invertebrate Palaeontology* (Bivalvia), by redescription of published genera and super-generic taxa in China from 1927 to 2007.

Evolutionary Treasures

Almatium gusevi
古氏阿拉木图虫
晚三叠世，新疆吐鲁番

Late Triassic; Turpan, Xinjiang
NIGPAS Collection #84840

Montlivaltia bangoinensis
班戈高壁珊瑚
早白垩世，西藏班戈

Early Cretaceous; Baingoin, Tibet
NIGPAS Collection #65771

Belemnopsis himalayansis
喜马拉雅似箭石
晚侏罗世—早白垩世，西藏江孜

Late Jurassic — Early Cretaceous; Gyangzê, Tibet
NIGPAS Collection #58801

箭石 / Belemnites

箭石是已灭绝的海洋头足动物（软体动物门），起源于泥盆纪，灭绝于白垩纪末；它类似乌贼，具矿化内骨骼，化石（鞘）个体大小一般为4~12 cm。其化石多见于中生代地层中，是侏罗纪、白垩纪海相地层的标准化石。中国学者杨遵义等发现并研究过中国西藏中生代和华南、新疆二叠纪箭石化石。

Belemnites are extinct, squid-like marine cephalopods, age ranging from Devonian to Cretaceous. They have mineralized internal hard parts preservable in the fossil record, with preserved fossil size range of 4–12 cm. Commonly found in Mesozoic strata, they are Jurassic and Cretaceous index fossils for dating marine sediments. Chinese scholars including Yang Zunyi et al. discovered and studied belemnite fossils from the Mesozoic of Tibet and the Permian of South China and Xinjiang.

Evolutionary Treasures

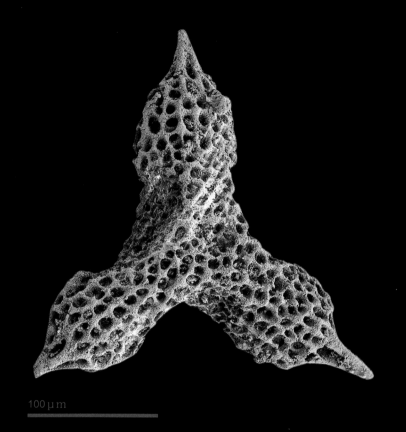

Paronaella muelleri
米勒帕氏虫
晚侏罗世—早白垩世，西藏江孜

Late Jurassic–Early Cretaceous; Gyangzê, Tibet
NIGPAS Collection #163027

100 μm

放射虫 / Radiolaria

放射虫是原生生物，为单细胞，绝大多数具有微小硅质骨骼，数十至数百微米大小。它在寒武纪至现代海洋中广泛分布，其化石丰富，是造山带地层划分对比的重要依据。20 世纪 70 年代，南京古生物所盛金章开辟了中国放射虫化石研究领域。20 世纪 80 年代以来，盛金章等一批研究者对中国西藏、新疆、东北和华南等地区的构造带和稳定沉积区的放射虫古生物化石及地层进行了广泛调查和研究，取得了一批重要研究成果。2009 年，南京古生物所科学家作为国际放射虫学家协会主席，成功组织了第 12 届国际放射虫学术年会。

100 μm

Pseudodictyomitra conicostriata
圆锥条纹假网冠虫
晚侏罗世—早白垩世，西藏浪卡子

Late Jurassic–Early Cretaceous; Nagarzê, Tibet
NIGPAS Colleciton #163000

100 μm

Archaeodictyomitra baergangensis
巴尔冈古网冠虫
早白垩世，西藏浪卡子

Early Cretaceous; Nagarzê, Tibet
NIGPAS Collection #162977

Radiolaria is a group of single-cell protistans, commonly with a siliceous shell, normally about 50 to 200 μm in size. They widely occur in worldwide oceans from Cambrian to the present. Radiolarian fossils are very important in division and correlation of orogenic stratigraphy. NIGPAS scientist Sheng Jinzhang pioneered in radiolarian micropalaeontology during the 1970s. Since the 1980s, a group of Chinese radiolarian specialists including Sheng Jinzhang et al. have conducted extensive investigations of orogenic strata in Tibet, Xinjiang and eastern Heilongjiang as well as stable sedimentary basins in South China and South China Sea, establishing the taxonomic and biostratigraphic framework. In 2009, NIGPAS micropalaeontologist successfully organized and chaired the 12th International Congress of Radiolaria in Nanjing.

恐龙模型（南京古生物博物馆展品）
Dinosaur models displayed in
Nanjing Museum of Palaeontology

前(front): *Tuojiangosaurus multispinus*
多棘沱江龙
后(rear): *Monolophosaurus jiangi*
江氏单脊龙

恐龙 / *Dinosauria*

恐龙是一度称霸地球、高度多样化的中生代史前动物，出现于大约 2.3 亿年前（三叠纪）。它在侏罗—白垩纪时主导陆地生态系统，于 6600 万年前灭绝，在地球上生存了 1.65 亿年之久。恐龙可依据腰带（骨盆）类型划分为蜥臀类和鸟臀类，前者又可区分出植食性的蜥脚类和肉食性的兽脚类。近年大量化石研究表明，鸟类起源于侏罗纪兽脚类恐龙祖先。因此，从分支系统学看，鸟类属于"带羽毛的恐龙"，是中生代恐龙的后代。

恐龙的学名 *Dinosauria*（拉丁文），意为"恐怖的蜥蜴"，但事实上，恐龙不属于任何现生爬行动物类群，更不是蜥蜴类。它们与现代爬行动物类群的显著差异是，恐龙类的四肢并不伸展呈坐躺姿态，而是呈站立姿态。研究表明，恐龙可能不是冷血动物，它们的体温可以调节。有些中生代爬行动物被公众当成恐龙，包括翼龙、鱼龙、蛇颈龙、沧龙、盘龙类，但是在分类学上它们不属于恐龙类。

中国是世界上最重要的恐龙化石产地之一。恐龙骨骼化石、恐龙遗迹及恐龙蛋化石产地遍布全国，极为丰富，已发现命名了近 200 余种，位居全球第一。

Dinosauria is a highly diverse group of prehistorical animials dominating the Mesozoic terrestrial environment. Dinosaurs first appeared during the Middle to Late Triassic times about 230 Ma, became the dominating animals on the land during Jurassic and Cretaceous periods, and went extinct at the end of Cretaceous about 66 Ma. *Dinosauria* as a clade is divided into two major groups, Saurischia and Ornithischia, based on the pelvis structure, the former including the herbivorous sauropods and carnivorous theropods. Recent studies show that birds were derived from a Jurassic theropod ancestor; thus, birds are considered "feathered dinosaurs", or the descendants of Mesozoic dinosaurs.

The Latin word *Dinosauria* means "horrible lizards", but dinosaurs are not closely related with lizards or other modern reptilian groups. Dinosaurs do not have typical modern reptilian features such as sprawling legs and cold-blooded; rather, dinosaur legs were generally upright in posture and some dinosaurs were likely active animals with mechanism of thermoregulation, according to recent studies.

China is one of the most important dinosaur-producing countries in the world, yielding dinosaur bones, eggs and trace fossils excavated throughout the region, with nearly 200 species of dinosaurs formally described and nominated.

Dinosaur eggs
恐龙蛋
晚白垩世，河南
南京古生物博物馆陈列标本

Late Cretaceous; Henan
Specimen displayed in Nanjing
Museum of Palaeontology

Evolutionary Treasures

Bernoullia zeilleri
蔡耶贝尔瑙蕨
晚三叠世，宁夏固原

Late Triassic; Guyuan, Ningxia,
NIGPAS Collection #PB2568

Danaeopsis fecunda
多实拟丹尼蕨
晚三叠世，新疆克拉玛依

Late Triassic; Karamay, Xinjiang
NIGPAS Collection #PB2581

Evolutionary Treasures

Neocalamites carcinoides
蟹形新芦木
晚三叠世，陕西延长

Late Triassic; Yanchang, Shaanxi
NIGPAS Collection #PB2257

Pachypteris lepingensis
乐平厚羊齿
三叠纪，江西乐平

Triassic; Leping, Jiangxi
NIGPAS Collection #PB4649

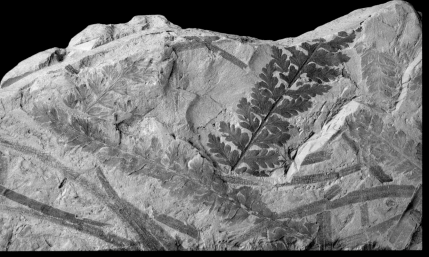

Coniopteris gansuensis
甘肃锥叶蕨
侏罗纪,甘肃张掖

Jurassic; Zhangye, Gansu
NIGPAS Collection #PB17056

Coniopteris burejensis
布列亚锥叶蕨
侏罗纪，青海柴达木

Jurassic; Qaidam, Qinghai
NIGPAS Collection #PB2637

Baiera furcata
叉状拜拉
侏罗纪，湖北秭归

Jurassic; Zigui, Hubei
NIGPAS Collection #PB940

Evolutionary Treasures

Ptilophyllum pectin
栉形毛羽叶
侏罗纪，湖北秭归

Jurassic; Zigui, Hubei
NIGPAS Collection #PB958

Sphenobaiera longifolia
长叶楔拜拉
早—中侏罗世，青海天峻

Early–Middle Jurassic; Tianjun, Qinghai
NIGPAS Collection #PB6401

银杏 / *Ginkgo biloba*

银杏是当今地球上最古老的树种之一，现存一科一属一种，属于裸子植物银杏目，其天然种群仅在中国分布。银杏类植物出现于二叠纪，可能源于种子蕨类祖先，于三叠纪晚期开始蓬勃发展，侏罗纪、白垩纪进入鼎盛时期。银杏属最早出现在侏罗纪早期，于侏罗纪中期至白垩纪早期高度分化，而在白垩纪末生物大灭绝事件后，仅有少数种类幸存。上新世以后，由于第四纪冰川的影响，银杏类在全球消失，仅剩一种残存在华中地区，成为植物"活化石"。南京古生物所周志炎院士等在河南义马中侏罗世地层中发现的义马果（*Yimaia*）代表了早期银杏家族中一个独立的演化支；我国早白垩世地层中发现的无柄银杏（*Ginkgo apodes*）代表了侏罗纪银杏和现生银杏之间的演化过渡类型。

Evolutionary Treasures

Ginkgo apodes
无柄银杏
早白垩世，辽宁义县

Early Cretaceous; Yixian, Liaoning
NIGPAS Collection #PB19881

Ginkgoites sibirica
西伯利亚似银杏
早—中侏罗世，青海天峻

Early–Middle Jurassic; Tianjun, Qinghai
NIGPAS Collections #PB6398

Ginkgo biloba is one of the longest living tree species on the Earth today, belonging to Subphylum Gymnospermae, Order Ginkgoales. This extant species is the only surviving member of the order. Ginkgoales first appeared during Permian times, probably evolved from the seed ferns, and diversified during Late Triassic, highly diversified during Jurassic and Cretaceous periods, with few species surviving the end-Cretaceous mass extinction event. Due to the Quaternary global glaciation after Pliocene times, ginkgos disappeared nearly worldwide, except for a small area in central China, where one relict species *Ginkgo biloba* has survived the glaciations and has been known as a "living fossil". It is noteworthy that the Middle Jurassic ginkgoalean *Yimaia*, discovered by NIGPAS palaeobotanist Zhou Zhiyan and his associates from the Yima Coal Mine, Henan, represents a separate branch in the ginkgoalean family, and the Early Cretaceous species *Ginkgo apodes*, also found in China, is considered a transition between the Jurassic and the living ginkgos.

"热河生物群" /"Jehol Biota"

热河生物群是白垩纪早期（早白垩世，1.45 亿~1.0 亿年前）一个极为繁盛的陆地生物群，分布于中国北方，蒙古、朝鲜等地。中国辽西是研究"热河生物群"的经典地区，主要产出地层为义县组和九佛堂组，跨越 1 亿多年地球历史，以"带毛的恐龙"、"世上最早的花"、"食恐龙的哺乳动物"、"长四个翅膀的恐龙"等发现而享誉世界。热河生物群的生活环境属于湿地和湖泊群，因此，保存下来的大量精美化石代表特异埋藏作用，可能与大量火山灰沉积有关。

1962 年，南京古生物所顾知微首先提出"热河生物群"的概念，以狼鳍鱼、三尾类蜉蝣和东方叶肢介三类常见化石组合为特征。20 世纪 90 年代以来，"热河生物群"研究取得快速进展，中国辽西也因此成为研究鸟类起源，以及翼龙、哺乳动物、早期被子植物和昆虫等生物类群进化的圣地。

"热河生物群"生态复原图(南京古生物博物馆)

Artist's reconstruction of palaeoecology of "Jehol Biota" in Nanjing Museum of Palaeontology

Jehol Biota is a very rich and exceptionally preserved Early Cretaceous terrestrial biota, with a distribution covering North China, Mongolia and the Korean Peninsula. West Liaoning (northeast China) is the classic area for studying the Jehol Biota, where the major fossil-yielding strata are the Yixian Formation and the Jiufotang Formation, spanning more than 100 million years. It is famous for the discoveries of "the feathered dinosaurs" "the earliest flowers" "the dinosaur-eating mammals" "the four-winged dinosaurs", etc. The palaeo-environment was characteristically wetlands and lakes, and the exceptional fossil preservation is most likely due to rapid volcanic ash falls.

In 1962, NIGPAS palaeontologist Gu Zhiwei proposed the concept of "the Jehol Biota" characterized by the common fossil assemblage of abundant *Lycoptera davidi* (fish), *Ephemeropsis trisetalis* (insect) and *Eosestheria* (conchostracan). Since the 1990s, studies on the Jehol Biota have advanced tremendously, especially for the origin and diversification of angiosperms, insects, birds, pterosaurs and mammals.

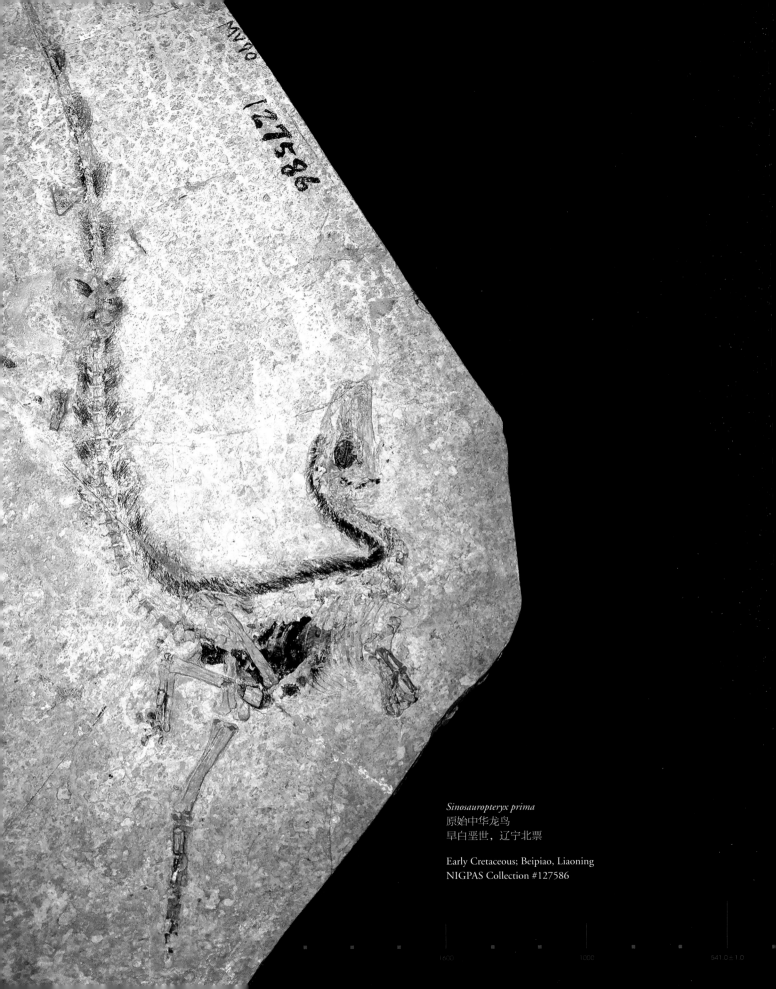

Sinosauropteryx prima
原始中华龙鸟
早白垩世，辽宁北票

Early Cretaceous; Beipiao, Liaoning
NIGPAS Collection #127586

中华龙鸟 / *Sinosauropteryx*

中华龙鸟,又名"中华鸟龙",属于美颌龙类,其化石产于辽西义县组(早白垩世),是"热河生物群"的著名成员。1996年,首次报道的中华龙鸟开始被认为是原始鸟类,后经南京古生物所古生物学家及国际合作团队重新研究,将其确定为小型食肉兽脚类恐龙。中华龙鸟尾部特别长,前肢很短,体长可达1 m多。其羽毛上保存的黑色素体表明,原始羽毛的颜色带有红色和浅色相间的条带。中华龙鸟体腔内含物显示,其食物中有蜥蜴和哺乳动物等;体内保存的卵的特征表明,它的生殖方式与其他兽脚类一样,是成对卵生的。

Sinosauropteryx is a group of small carnivorous dinosaurs, belonging to the therapod comsognathids, from the Lower Cretaceous Yixian Formation, West Liaoning, a star of the Jehol Biota. When first found in 1996, the fossil was considered to a primitive bird, but a reexamination by NIGPAS palaeontologist Chen Piji and his international team determined its dinosaur affiliation. *Sinosauropteryx* has a very long tail and short arms, with a total body length of more than 1 m. Studies show that there are preserved melanosomes in the feather, indicating the original coloration of the feather with reddish and lighter bands. Relict contents found in the body cavity indicate the dinosaur's diet included lizards and mammals, while unlaid eggs preserved in the body suggest that *Sinosauropteryx* laid eggs in pairs like other therapod dinosaurs.

Liaoxiornis delicatus
娇小辽西鸟
早白垩世,辽宁凌源

Early Cretaceous; Lingyuan, Liaoning
NIGPAS Collection #130723

孔子鸟 / *Confuciusornis*

孔子鸟是原始鸟类的一个属，距今约 1.25 亿~1.20 亿年，其化石产于辽西下白垩统义县组和九佛堂组中。孔子鸟与现代鸟类一样都是喙部无齿，是目前已知最早拥有无齿角质喙部的鸟类；体形如鸽子大小，翼展约 0.7 m，体重为 0.2~1.5 kg。孔子鸟的属名源于中国春秋时期著名哲学家孔子，由著名古鸟类专家侯连海领导的团队于 1995 年首次描述。

Evolutionary Treasures

Confuciusornis is a genus of primitive fossil birds, discovered from the Early Cretaceous strata of the Yixian and Jiufotang Formations in West Liaoning, dated about 125–120 Ma. Like modern birds, Confuciusornis had a toothless beak, representing the earliest occurrence of such characteristic of birds lacking teeth with the horny beak. The primitive bird was about the size of a modern pigeon, with a wingspan of approximately 0.7 m and a weight of 0.2 to 1.5 kg. The genus name was derived from Confucius — the famous philosopher of the Spring and Autumn Period in ancient China, first reported by fossil bird specialist Prof. Hou Lianhai and his team in 1995.

Confuciusornis sanctus
圣贤孔子鸟
早白垩世，辽宁北票

Early Cretaceous; Beipiao, Liaoning
NIGPAS Collection #139379

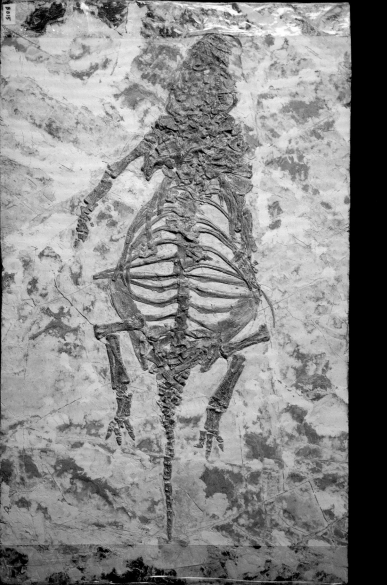

Ankylosaurus sp.
甲龙（未定种）
早白垩世，辽西

Early Cretaceous; West Liaoning

Sinohydrosaurus lingyuanensis
凌源中国水生蜥
早白垩世，辽宁凌源

Early Cretaceous; Lingyuan, Lia
NIGPAS Collection #MV79

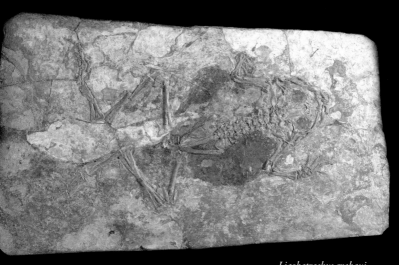

Liaobatrachus grabaui
葛氏辽蟾
早白垩世，辽宁北票

Early Cretaceous; Beipiao, Liao
NIGPAS Collection #MV77

Evolutionary Treasures

Palaeocambarus licenti
桑氏古蝲蛄
早白垩世，辽宁凌源

Early Cretaceous; Lingyuan, Liaoning
NIGPAS Collection #SYB(2000)-68

Cricoidoscelosus aethus
奇异环足虾
早白垩世，辽宁凌源

Early Cretaceous; Lingyuan, Liaoning
NIGPAS Collection #SYB(2000)-71

Evolutionary Treasures

Gymnogrammitites ruffordioides
鲁福德蕨型似雨蕨
早白垩世，辽宁北票

Early Cretaceous; Beipiao, Liaoning
NIGPAS Collection #PB19020

Rehezamites anisolobus
不等裂热河似查米亚
早白垩世，辽宁北票

Early Cretaceous; Beipiao, Liaoning
NIGPAS Collection #PB18265

左图(left): *Problematospermum ovale*
卵形毛籽
早白垩世，辽宁北票

Early Cretaceous; Beipiao, Liaoning
NIGPAS Collection #PB19191

右图(right): *Liaoxia chenii*
陈氏辽西草
早白垩世，辽宁北票

Early Cretaceous; Bepiao, Liaoning
NIGPAS Collection #PB17800

Parapetala liaoningensis
辽宁类古蜓
早白垩世,辽西

Early Cretaceous; West Liaoning
NIGPAS Collection #133694

"燕辽生物群" / "Yanliao Biota"

"燕辽生物群"是分布于辽西、冀北和内蒙古东南部的中—晚侏罗世陆生生物化石群,相当于"道虎沟生物群",主要产于海防沟组和蓝旗组(河北称九龙山组和髫髻山组),包含丰富的无脊椎动物(叶肢介类、双壳类、腹足类、昆虫类、蜘蛛类等)化石、脊椎动物(长羽毛兽脚类恐龙、原始哺乳类、翼龙类、蝾螈类、鱼类等)化石和植物化石。"燕辽生物群"主要保存于湖相沉积中。"燕辽生物群"与早白垩世的"热河生物群"有相似之处,但其多样性较低且生物组合差别明显。

Nodalula dalinghensis
大凌河结翅蜓
早白垩世，辽西

Early Cretaceous; West Liaoning
NIGPAS Collection #123517

Yanliao Biota is a Middle–Late Jurassic land biota distributed in Southeast Inner Mongolia, West Liaoning, and northern Hebei; sometimes also referred to as the "Daohugou Biota", mainly yielded from the Haifanggou and Lanqi formations (or the Jiulongshan and Tiaojishan formations in Hebei). Yanliao Biota includes very abundant invertebrate fossils (such as conchostracans, bivalves, gastropods, insects and spiders), vertebrate fossils (e.g., fishes, salamanders, lizards, pterosaurs, therapods and mammals) and plant fossils. The abundant and exceptionally preserved fossils are mainly found in fine volcanic ash deposits of lacustrine settings. Researchers found similarities between the Yanliao Biota and the Early Cretaceous Jehol Biota from the same region, but the former differs in relatively lower diversity and distinct taxonomic composition.

Evolutionary Treasures

Mesocupes angustilabialis
窄唇中生长扁甲
中侏罗世，内蒙古宁城

Middle Jurassic; Ningcheng, Inner Mongolia
NIGPAS Collection #142145

Daohugounectes primitivus
原始道虎沟桨甲
中侏罗世，内蒙古宁城

Middle Jurassic; Ningcheng, Inner Mongolia
NIGPAS Collection #149560

Coleoptera
鞘翅目昆虫
中侏罗世，内蒙古宁城

Middle Jurassic; Ningcheng, Inner Mongolia
NIGPAS Collection #149371

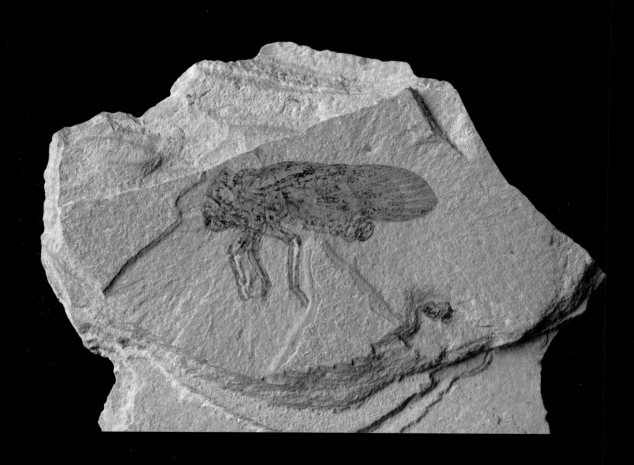

Shuraboprosbole minuta
娇小舒拉布螽蝉
中侏罗世，内蒙古宁城

Middle Jurassic; Ningcheng, Inner Mongolia
NIGPAS Collection #149376

Beipiaoserphus elegans
华美北票细蜂
早白垩世，辽宁北票

Early Cretaceous; Beipiao, Liaoning
NIGPAS Collection #132048

Praeaulacus daohugouensis
道虎沟原举腹蜂
中侏罗世，内蒙古宁城

Middle Jurassic; Ningcheng, Inner Mongolia
NIGPAS Collection #148217

Evolutionary Treasures

Silphid
葬甲科昆虫
中侏罗世，内蒙古宁城

Middle Jurassic; Ningcheng, Inner Mongolia
NIGPAS Collection #156144a

Tyrannopsylla beipiaoensis
北票霸王蚤
早白垩世，辽宁北票

Early Cretaceous; Beipiao, Liaoning
NIGPAS Collection #154249a

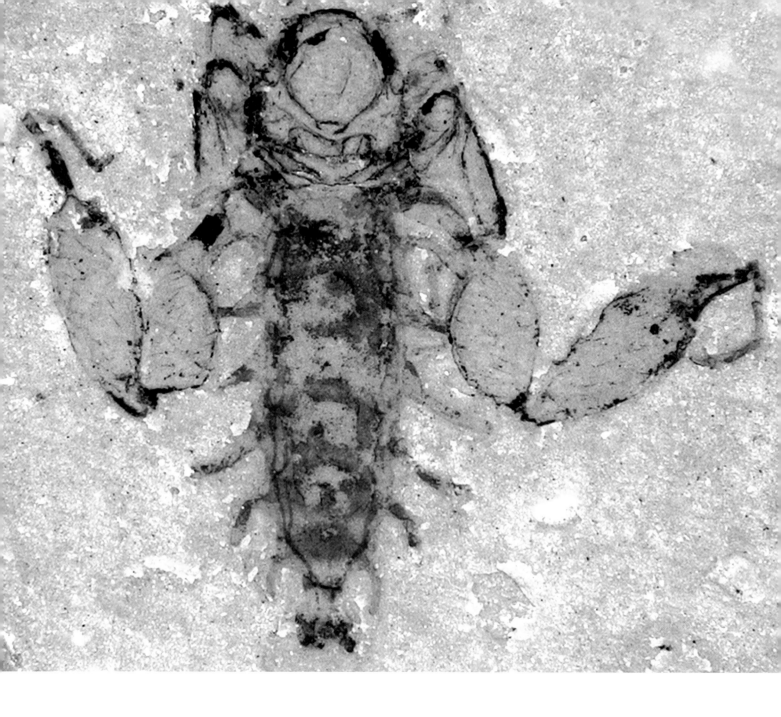

Strashila daohugouensis
道虎沟恐怖虫
中侏罗世，内蒙古宁城

Middle Jurassic; Ningcheng, Inner Mongolia
NIGPAS Collection #155017

Evolutionary Treasures

Callianthus dilae
迪拉丽花
早白垩世，辽宁北票

Early Cretaceous; Beipiao, Liaoning
NIGPAS Collection #PB21047

被子植物的起源 / The Origin of Angiosperms

被子植物也称"有花植物"，是当今世界上多样性最高的陆生植物。150多年前，进化生物学奠基人达尔文，曾因被子植物化石突然在白垩纪地层中大量出现，却又找不到它们的祖先类群和早期演化的线索，因而称之为"讨厌的谜团"。100多年来，全世界古植物学家和植物学家都在孜孜不倦地为这一"讨厌的谜团"寻求答案。

Angiosperms are flowering plants that represent the most diverse group of terrestrial plants in modern world. As early as 150 years ago, Charles Darwin was puzzled at the sudden abundant appearance of angiosperm fossils in the Cretaceous strata without any clue of their ancestral links, for which he called "an abominable mystery". Since then, botanists and palaeobotanists worldwide have been searching for an answer to the "abominable mystery".

Evolutionary Treasures

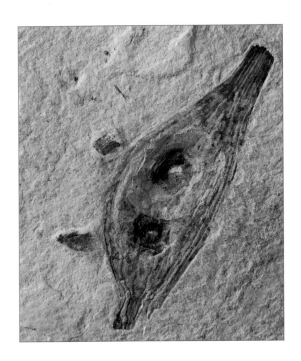

Liaoningfructus ascidiatus
瓶状辽宁果
早白垩世，辽宁北票

Early Cretaceous; Beipiao, Liaoning
NIGPAS Collection #PB21405

近年来，南京古生物所等机构的中国科学家在"热河生物群"等化石群中惊人地发现了早期被子植物化石。辽宁古果（*Archaefructus liaoningensis*）、中华古果（*A. sinensis*）、十字里海果（*Hyrcantha decussata*）、中华施迈斯内果（*Schmeissneria sinensis*）、中华星学花（*Xingxueanthus sinensis*）、迪拉丽花（*Callianthus dilae*）和瓶状辽宁果（*Liaoningfructus ascidiatus*）等一系列早期被子植物或具有被子植物特征的化石的发现，丰富了我们对被子植物起源和早期演化的认识。

In recent years, NIGPAS palaeontologists and those from other institutions in China together with their international collaborators made surprising discoveries of early angiosperm fossils in the Jehol Biota and other fossil Lagerstätten in West Liaoning, Northeast China. The publication and characterization of the exceptionally preserved fossil species *Archaefructus liaoningensis*, *A.sinensis*, *Hyrcantha decussata*, *Schmeissneria sinensis*, *Xingxueanthus sinensis*, *Callianthus dilae*, *Liaoningfructus ascidiatus*, assigned to early angiosperms or with angiospermous characteristics have substantially advanced our understanding of the origin and early evolution of angiosperms.

Archaefructus liaoningensis
辽宁古果
早白垩世,辽宁凌源

Early Cretaceou; Lingyuan, Liaoning
NIGPAS Collection #PB19283

新生代

Cenozoic:
Era of New Life after Ammonites and Dinosaurs

Phanerozoic >

Evolutionary Treasures

新生代是显生宙的最后一个地质时代（继古生代、中生代之后），始于6600万年前。如果说中生代是"恐龙与裸子植物时代"的话，新生代便是"哺乳动物与被子植物时代"。继恐龙等大型爬行动物在白垩纪末灭绝之后，哺乳动物在新生代快速演化，成为陆地的主宰。因干旱而形成的大片稀树草原也是新生代陆地生态中的一个新亮点，它们是包括人类在内的现今陆生动物物种的起源与演化的摇篮。新生代的无脊椎动物与其现生类群非常接近，演化进程发生在种、属的水平上。双壳类、腹足类、介形类、海洋单细胞生物和植物孢粉等化石，在确定地层时代和沉积环境等方面具有重要意义。昆虫与被子植物的互助与共同繁盛是新生代生态系统的主旋律，二者成为地球上物种数量最多的生物。

新生代分为3个纪（古近纪、新近纪、第四纪）7个世（古新世、始新世、渐新世、中新世、上新世、更新世、全新世），而以往常用的"第三纪"已被正式分解为古近纪和新近纪。由于第四纪以古人类的出现和人类演化为特征，因而有的学者称之为"人类纪"；最近有学者提出的"人类世"，指的是新石器时期农耕开始（12000多年前）以来，人类活动对于地球环境和生物多样性产生显著影响的地球时期。

新生代的全球板块构造运动剧烈，造就了当前的海陆格局，其主要特征是南方冈瓦纳大陆继续分离；印度次大陆北漂与欧亚板块碰撞，形成雄伟的喜马拉雅山脉和缓缓隆升的青藏高原；阿拉伯半岛与欧亚板块碰撞形成扎格罗斯山脉；古地中海闭合；当今世界上的高山——安第斯山、落基山、阿尔卑斯山也都在新生代崛起；澳洲脱离南极洲后北漂，并与东南亚板块碰撞；南美洲与南极洲分离，并与北美洲在巴拿马地峡处碰接；环南极海形成；大西洋、北冰洋等大片海域扩张。

Archean >   Proterozoic >

Eoarchean	Palaeoarchean	Mesoarchean	Neoarchean	Palaeoproterozoic				Mesoproterozoic				Neoproterozoic
				Siderian	Rhyacian	Orosirian	Statherian	Calymmian	Ectasian	Stenian	Tonian	Cryogenian
4000 3600	3200	2800	2500	2300	2050	1800	1600	1400	1200	1000	720	

Time (Ma)

Cenozoic is the last one of the three major geological eras of the Phanerozoic Eon, following the Palaeozoic and Mesozoic. It began 66 Ma. If you refer to the Mesozoic as the "Age of Dinosaurs and Gymnosperms", Cenozoic is sometimes called the "Age of Mammals and Angiosperms". As dinosaurs and many other giant reptiles disappeared at the end-Cretaceous mass extinction event, mammals experienced rapid diversification and climbed to the top part in the food pyramid during the Cenozoic times. Due to climate change, large expanses of savannas developed, becoming the likely cradles for the origin and earlier evolution of many modern animals including our own group — the genus *Homo*, another characteristic feature of the Cenozoic life. Invertebrate animals of the Cenozoic are very close to the modern groups and their evolution occurred at the generic and species levels. Fossil groups such as bivalves, gastropods, ostracods, protists, pollen and spores of the plants play very important roles in determining the stratigraphic ages and sedimentary environment for the Cenozoic. The mutual dependence and great diversification of insects and angiosperms, both being the most diverse life forms on the Earth, also features the great life evolution of the Cenozoic.

The Cenozoic Era includes three periods (Palaeogene, Neogene and Quaternary) and seven epochs (Palaeocene, Eocene, Oligocene, Miocene, Pliocene, Pleistocene and Holocene), while the old period name "Tertiary" has been dissolved to become Palaeogene and Neogene. Although the term "Anthropogene" was used for the Quaternary Period for the hominoid evolution, recent proposals for Anthropocene refers to the time interval of significant anthropogenic influence to the earth environment and biodiversity since the rise of agriculture about 12,000 years ago.

The Cenozoic tectonic activities are dramatic, creating the current configuration of continents and oceans. Parts of the earlier supercontinents Pangea continued to separate and moved away from each other: the Indian subcontinent continued its northward journey and collided with Eurasia pushing up the mighty Himalayas and the Qinghai–Tibet Plateau; the Arabian Peninsula collided with Eurasia forming the Zagros Mountains; the Tethys closed. Today's major mountain ranges on the Earth were products of the Cenozoic tectonics: the Andes, Rockies and Alps. Australia detached from the Antarctica and moved northward to collide with Southeast Asia; South America separated with Antarctica and met North America at the Panama Isthmus; the Circum-Antarctic Current formed; large oceanic areas were created in the Atlantic, Artic regions and elsewhere.

Phanerozoic >

Evolutionary Treasures

伴随海陆板块格局的演化，新生代全球气候发生了剧变。与中生代晚期的湿、热气候相比，新生代气候总体呈现变冷、变干的趋势，主要原因是南极环流的形成，有效地冷却了全球海洋，最终导致第四纪全球冰川的形成（目前地球处于第四纪冰期中的间冰期，即全新世）。新生代生命演化与全球构造演变及气候变化密切相关。

The tectonic evolution greatly influenced the Cenozoic climatic change, featuring a general cooling and drying trend, finally leading to the Quaternary Glaciation, for which researchers proposed, as a primary mechanism, the formation of the Circum-Antarctic Current, which effectively cooled the world oceans; we are currently at an inter-glacial period within the Quaternary Glaciation. The Cenozoic life evolution may be closely tied with tectonic and climatic changes.

"抚顺琥珀生物群" / "Funshun Amber Biota"

抚顺琥珀主要产于辽宁抚顺矿区始新世古城子组的煤系地层中。南京古生物所研究人员系统研究了"抚顺琥珀生物群"中的化石，发现了大量多样性很高的节肢动物、植物化石（水杉为主）和微体化石。这是世界上种类最丰富的琥珀生物群之一。它距今 5000 多万年。

琥珀是古植物树脂的化石，是地质时期植物分泌的树脂在沉积物中经过漫长的成岩作用（在一定的压力和温度下）而形成的。琥珀为一些古动物和古植物提供了极好的埋藏条件，是一种独特的化石宝库。已知最早的琥珀化石产于大约 3.2 亿年前的石炭纪晚期煤层中，琥珀在石炭纪以后的陆相沉积物中更为丰富。研究者已从琥珀中发现昆虫、蜘蛛、螨虫、环节类、蛙类、花、果实、毛发、微体生物以及生物器官等。

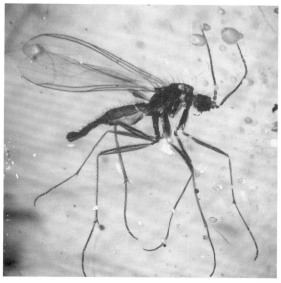 

Sciarid fly
眼蕈蚊
始新世，辽宁抚顺

Eocene; Fushun, Liaoning

Parasitoid wasp
茧蜂
始新世，辽宁抚顺

Eocene; Fushun, Liaoning

Fushun ambers are found in the Eocene coal-bearing strata of the Guchengzi Formation in the Fushun Coal Mines, Liaoning, NE China. NIGPAS researchers found that the Fushun Amber Biota is among most diverse amber Lagerstätten worldwide, with very abundant and highly diverse fossil arthropods together with abundant plant and microfossils; the Biota is dated over 50 Ma.

Ambers are fossilized resins of ancient plants; they were formed through prolonged diagenetic process of the resin-embedding sediments under high temperature and pressure. Ambers provide exceptional taphonomic conditions for many ancient animals and plants that would otherwise not have survived for discovery, thus becoming a unique fossil Lagerstätten for palaeobiological research. The earliest known ambers are from coal sediments of late Carboniferous about 320 Ma, while post-Carboniferous ambers occur more abundantly in terrestrial sediments. Researchers have found abundant insects, spiders, mites, annelids, frogs, flowers and fruits, hairs, microfossils and animal organs from ambers worldwide.

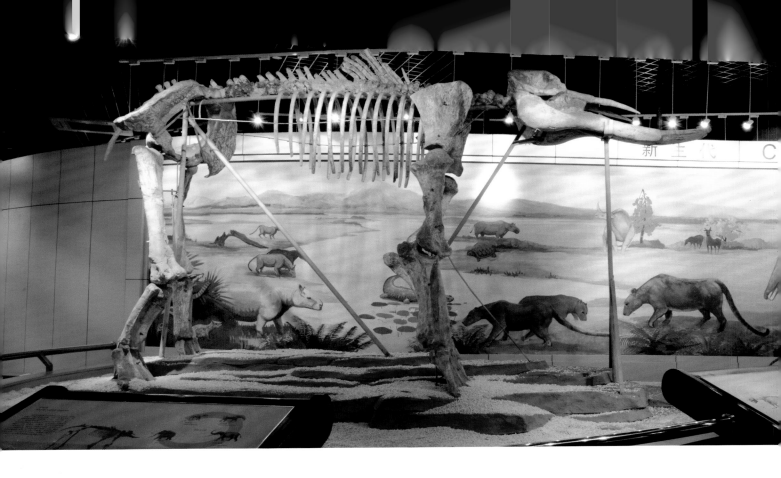

"和政动物群" /"Hezheng Faunas"

"和政动物群"是著名的新生代生物化石群,产于甘肃临夏和政县及附近地区的新近纪和第四纪地层中,是整个欧亚大陆最丰富的哺乳动物化石群之一,伴生其他动植物(包括龟类、介形类、孢粉等)化石。"和政动物群"包括著名的铲齿象动物群(中新世中期,距今约 1300 万年)、三趾马动物群(中新世晚期,距今约 1000 万年)、巨犀动物群(晚渐新世,距今约 3000 万年)和真马动物群(早更新世,距今约 200 万年)。

"和政动物群"所在的临夏盆地位于青藏高原和黄土高原交界地带,发育了一套从渐新世至更新世的连续沉积地层,因此,"和政动物群"的演化见证了青藏高原的隆升过程,反映了该地区从亚热带森林密布,到季节分明的干旱草原,进而随着青藏高原的缓缓隆起,逐渐演变为干冷、生态日益恶化的第四纪黄土沉积环境的过程。

新生代动物化石群展示（南京古生物博物馆）

Exhibition of Cenozoic Biota in the Nanjing Museum of Palaeontology

The Hezheng Faunas are famous Cenozoic faunas, discovered from the Neogene to Quaternary strata of Hezheng and neighboring counties of Linxia, Gansu, regarded as one of the most diverse mammal faunas on the Eurasian continent, with other co-occurring animals and plants (such as turtles, ostracods and spores and pollen). The Hezheng Faunas contain the most well-known Platybelodon Fauna (middle Miocene, about 13 Ma) and Hipparion Fauna (late Miocene, about 10 Ma), as well as Dzungariotherium Fauna (late Oligocene, about 30 Ma) and Equus Fauna (early Pleistocene, about 2 Ma).

The Linxia Basin, which produces the Hezheng Faunas, is located at the juncture of the Qinghai–Tibet Plateau and the Loess Plateau, and contains a continuous sequence of Oligocene to Pleistocene sediments; thus, the succession of Hezheng Faunas are witnesses of the rise and influence of the Qinghai–Tibet Plateau, indicating an environmental evolution from dense subtropical forests, to seasonal dry climatic steppes, and then to cold and dry Quaternary loess depositional environment.

Evolutionary Treasures

Testudo sp.
陆龟(未定种)
新近纪,甘肃

Neogene; Gansu

Indarctos zdanskyi
师氏印度熊
新近纪，甘肃

Neogene; Gansu
NIGPAS Collection #MV172

Amphimachairodus palanderi
巴氏剑齿虎
新近纪，甘肃

Neogene; Gansu
NIGPAS Collection #MV169

"山旺生物群" / "Shanwang Biota"

"山旺生物群"（中新世中期，距今约 1400 万年）产自山东临朐山旺盆地硅藻土地层，享有"万卷书"之称。山旺生物群化石主要有大量昆虫、鱼类、蜘蛛、两栖类、爬行类、鸟类及哺乳类等动物化石，以及苔藓、蕨类、裸子植物、被子植物及藻类等植物化石，种类之多、保存之完整为世界罕见。山旺地层形成于内陆湖泊沉积盆地，由泥岩、硅藻土岩、油页岩夹玄武岩组成，是我国东部沿海地区中新世中期的标准地层。

Shanwang Biota is preserved in the middle Miocene diatomite and related deposits of the Shanwang Basin in Linqu, Shandong. Because of the highly fossiliferous and finely laminated nature of the sediments in the Shanwang Basin, it is widely known as the "Myriad Archives". Major fossil groups include abundant animals such as insects, fish, spiders, amphibians, reptiles, birds and mammals as well as plant fossils such as mosses, ferns, gymnosperms, angiosperms and algae. The high abundance and diversity and superb preservation qualities of the biota are among the most remarkable worldwide. The sediments of the Shanwang Basin, mainly consisting of mudstones, diatomites and oil shales, were deposited in a continental lacustrine settings, and the sedimentary sequence has become the middle Miocene stratigraphic standard for the east coastal region of China.

Prunus miodavidiana
中新桃
中新世，山东山旺

Miocene; Shanwang, Shandong
NIGPAS Collection #No.9 (H)

Evolutionary Treasures

Plesiaceratherium gracile
细近无角犀
中新世，山东山旺

Miocene; Shanwang, Shandong

Tetrastigma shantungensis
山东崖爬藤
中新世，山东山旺

Miocene; Shanwang, Shandong
NIGPAS Collection #No.208 (H)

Evolutionary Treasures

Palaeomastodon sp.
古乳齿象（未定种）（臼齿）
渐新世，甘肃

Oligocene; Gansu

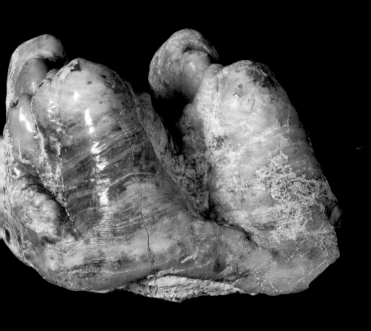

Elephas namadicus (3rd molar)
纳玛象（第三臼齿）
第四纪，江苏南京

Quaternary; Nanjing, Jiangsu

Quercus kodairae
小平栎
渐新世, 吉林龙井

Oligocene; Longjing, Jilin
NIGPAS Collection #PB14655

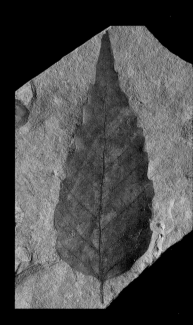

Fagus koraica
科瑞山毛榉
渐新世, 吉林龙井

Oligocene; Longjing, Jilin
NIGPAS Collection #PB14661

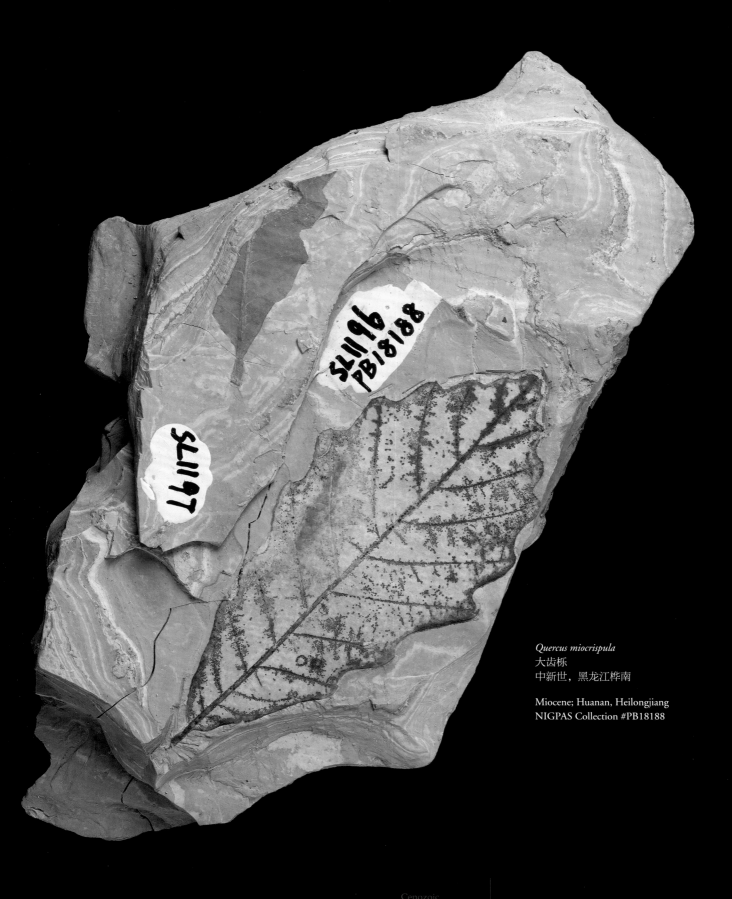

Quercus miocrispula
大齿栎
中新世，黑龙江桦南

Miocene; Huanan, Heilongjiang
NIGPAS Collection #PB18188

Evolutionary Treasures

Populus norini
变叶杨
上新世，青海柴达木

Pliocene; Qaidam, Qinghai
NIGPAS Collection #PB6141

Taxodium tinajorum
丁纳落羽杉
古新世，新疆阿勒泰

Palaeocene; Altay, Xinjiang
NIGPAS Collection #PB13901

Nothofagogolia multinervis
多脉假山毛榉
始新世，南极乔治王岛化石山

Eocene; King George Island, Antarctica
(Collected by Li Haomin)
NIGPAS Collection #PB20805

Evolutionary Treasures

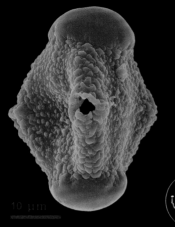

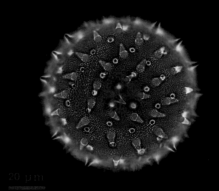

Sonneratia caseolaris
海桑
现代花粉，虚拟色

Modern pollen image with artificial coloration
Mao Limi Collection

Callirhoe involucrata
矮粟葵
现代花粉

Modern pollen image
Mao Limi Collection

孢粉 / Pollen & Spores

孢粉是孢子和花粉的简称，分别指孢子植物和种子植物的生殖细胞。孢粉学是研究孢子、花粉以及带有机质壁的微体化石（孢粉型化石）的一个学术分支，属于地学和植物学的一个交叉领域。由于孢子和花粉具有抵御降解的外壁，因此孢粉化石在前寒武纪至新生代地层中（尤其是陆相地层中）非常普遍，它们在地层学、古环境重建、地质与能源勘探以及早期生命演化（疑源类）和陆地生态环境的形成与演化探索中，扮演重要角色。南京古生物所以宋之琛、欧阳舒等为代表的几代科学家长期致力于中国各区域的孢粉化石探索和研究，在前寒武纪、古生代、中生代和新生代孢粉化石研究中，取得了一系列重要成果。

Land plants produce pollen and spores (reproductive cells) that are extremely resistant and widely preserved in sedimentary rocks. Palynology is an interdisciplinary research area between geosciences and botany, studying pollen, spores, and other organic-walled microfossils (palynomorphs) in the geological record from Precambrian to Cenozoic sediments. Palynological studies play an important role in stratigraphy, palaeoenviromental reconstruction, geological exploration for energy and other resources, studies of early life evolution and the origin of terrestrial ecosystem. NIGPAS scientists represented by Song Zhichen, Ouyang Shu et al. have been actively engaged in palynological investigations throughout the regions of China and made remarkable contributions to studies of Precambrian to Cenozoic strata.

Spencerisporites striatus
条纹斯潘塞孢
晚石炭—早二叠晚期

Late Carboniferous–Early Permian
Liu Feng Collection

100 μm

Hamamelis mollis
金缕梅
现代花粉

SEM image of modern pollen
Mao Limi Collection

10 μm

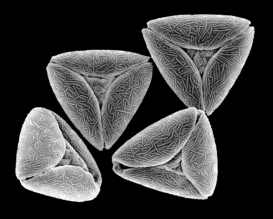

Nymphoides peltatum
荇菜
现代花粉

SEM image of modern pollen
Mao Limi Collection

20 μm

214 - 215

介形类 / Ostracodes

介形类是甲壳纲节肢动物，水生，个体微小（多数 0.2~30 mm），具有两瓣壳，壳质几丁质或钙质。其化石非常丰富，广泛分布于全球寒武纪至第四纪地层中，对地层的划分及对比具有重要意义。南京古生物所侯祐堂领导的介形类研究团队从 20 世纪 50 年代开始在中国开展介形类化石研究，致力于中国中、新生代非海相以及古生代海相介形类化石的开创性研究与探索，在油气地质、煤田地质、华南红层等领域中发挥了重要作用。

Ostracodes are small crustacean arthropods, living in aquatic (marine and terrestrial) environment, common size range of 0.2–30 mm, with a bivalve shell of chitinous or calcareous composition. Ostracodes are very abundant in the fossil record, with a worldwide distribution in Precambrian through Quaternary strata, playing a very important role in stratigraphic division and correlation. NIGPAS research team led by Hou Youtang initiated ostracod fossil studies during the 1950s, conducting extensive investigations of Chinese non-marine Mesozoic–Cenozoic rocks and marine Palaeozoic rocks, making significant contribution to petroleum and coal geology, and regional geological explorations.

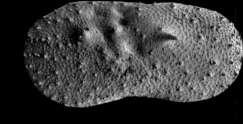

100 μm

Ilyocypris echinata
多刺土星介
更新世，青藏高原若尔盖盆地

Pleistocene; Zoigê Basin, Qinghai-Tibet Plateau
NIGPAS Collection #153634

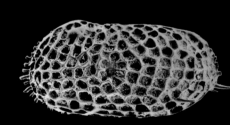

100 μm

Pistocythereis guangdongensis
广东纯艳花介
第四纪（海相），香港鲤鱼门

Quaternary (marine), Liyumen, Hong Kong
NIGPAS Collection #130003

钙质超微化石 / Calcareous Nannofossils

钙质超微化石是颗石藻（单细胞浮游藻类）的骨片，直径一般为 2~25 μm，藻体大小约 5~100 μm。钙质超微化石是一类重要的微体化石，沉积物中一般以孤立骨片保存，很少保存为完整球形个体。颗石藻化石广泛保存于海相沉积地层中，具有重要的生物地层学意义。颗石藻最早出现于晚三叠世卡尼期。颗石藻在现代海洋中也特别丰富，部分颗石藻有一定的温度适应范围，可分为暖水种、冷水种、广温种、窄温种，在古海洋研究中可发挥重要作用。南京古生物所钟石兰等在中国中新生代地层中对颗石藻化石进行了广泛的探索。

Calcareous nannofossils are skeletal plates of single-cell phytoplantonic algae known as coccolithophores. The coccolithophore cells are about 5–100 μm in size, and the plates referred to as coccoliths, which cover the cell surface, are about 2–25 μm. Coccoliths are widely and abundantly preserved in marine sediments. Calcareous nannofossils first appeared in the geological record from Late Triassic (Carnian) strata, playing a very important role as a biostratigraphic tool. Coccolithophores are still very abundant in modern oceans, with some species adapted to certain temperature range, as warm water species, cold water species, eurythermal or stenothermal species, useful in palaeoceanographic reconstruction. NIGPAS scientist Zhong Shilan et al. conducted extensive investigation on coccoliths in China.

Emiliania huxleyi
赫胥黎伊米莉亚藻
现代样品，中国南海

SEM image of modern sample; South China Sea

Lithastrinus floralis
花形腔石藻
晚白垩世，新疆乌恰

Late Cretaceous; Wuqia, Xinjiang

Discoaster bashibulakiensis
巴什布拉克盘星石藻
晚古新世，新疆乌恰

Late Palaeocene; Wuqia, Xinjiang

有孔虫属于单细胞原生生物，绝大多数生活在海洋中，具有钙质壳或胶结壳；一般个体小于 1 mm，有些可达数厘米或更大，例如新生代早期（古近纪）的货币虫和古生代晚期的䗴类有孔虫。有孔虫化石记录非常丰富，经常成为一些石灰岩的重要成分。有孔虫最早出现于寒武纪，晚古生代以来快速演化，其化石分布广泛，成为十分重要的生物地层学工具。南京古生物所首任所长李四光是我国（䗴类）有孔虫研究的早期代表，他创造了中文字"䗴"并创建了此类化石的鉴定标准。20 世纪 50 年代开始，研究所盛金章领导的有孔虫专业组对我国主要的含有孔虫化石的地层进行了广泛的调查和研究，尤其在石炭、二叠纪地层学和有孔虫系统学方面，做出了国际领先的成果。国内著名的有孔虫专家还有郝诒纯院士、郑守仪院士、汪品先院士等。

Foraminifera is a group of single-celled marine protists commonly with a calcareous or agglutinated shell (test). Test size is usually less than 1 mm, but some may reach a few centimeters or larger, such as the Palaeogene nummulites and Late Palaeozoic fusulinids. Foraminifers are very abundant in the fossil record, sometimes making a significant part of limestones. They first appeared during the Cambrian and became rapidly diversified and quick evolving from Late Palaeozoic onward, thus becoming an exceptional biostratigraphic tool for marine sediments. The NIGPAS founding director Li Siguang was a pioneering researcher of foraminiferal fossils in China; he coined the Chinese character for fusulinid foraminifers and proposed an identification standard of fusulinids. From the 1950s, NIGPAS scientist Sheng Jinzhang and his special foraminiferal group conducted extensive investigation of foraminifer-bearing strata in China, with outstanding contribution to the Carboniferous and Permian palaeontology and stratigraphy. Other Chinese

Uvigerina mediterranea
地中海葡萄虫

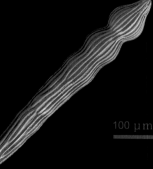

Dentalina albatrossi
信天翁齿形虫

Amphicoryna sublineata
近线状双棒虫

Globorotalia scitula
秀圆辐虫

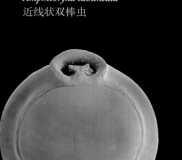

Pyrgo subsphaerica
亚球状双玦虫

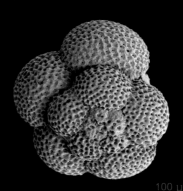

Neogloboquadrina dutertrei
杜氏新方球虫

Lagena striata
线纹瓶虫

第四纪有孔虫
大西洋（扫描电子显微影像）

Quaternary foraminifers
Atlantic (SEM images)
Li Baohua Collection

后 记

这本图集展示的化石标本，是大自然历经亿万年翻江倒海、冰火交融、海陆竞生后留给人们的，是岁月的沉淀，更是菁华的凝练。

人们发现的每一件化石标本，都对应着一个或一群在地球上生活过的生物，而化石标本代表的物种数量，甚至不及在地球上生活过的物种的百分之一；它们只是冰山一角、沧海一粟。因此，对于每一件化石标本而言，它们历尽沧桑，是见证地球远古生命世界的"幸存者"，无声地演绎着其中一幕又一幕"奇迹"般的故事，或为"伊甸园式"，或为"灾难式"。因而，将这些弥足珍贵的化石标本称为"菁华"，并不为过。

回溯至亿万年前，绝大多数生物体死亡之后，在各种自然营力的作用下，化为尘土，融入江河湖海，进入大自然的下一个循环。仅有极少数生物体死亡后，被裹挟到河流、湖泊或海洋的沉积物中，意外得到"庇护"，躲过动物的捕食，逃过微生物的降解，之后经历漫长的成岩过程和严酷的地质作用，宛如经历一次又一次的"炼狱"，其留存并被发现和研究的概率之小可想而知。

时至今日，这些幸存的化石，已经被人们用以重建地球生物宏大演化史，启示人们生物界演化的起起落落，揭示生物界存在的竞争与依存、分工与合作、继承与适应变化等复杂而巧妙的现象，展示生物与无机环境之间的互动与演进。这些是古生物专业研究者们孜孜探索的奥秘，也是化石爱好者们业余消遣的乐事。

如果将化石所处的沉积地层称为"万卷书"，化石便是记载地球远古生物世界的"密码"。研究者们循着地层的线索，探究远古生物的踪迹，就如同在"解码"一部大自然编写的"天书"。

地球上的生物（包括人类自己）是如何一步一步从远古走过来的？地球上的生物未来可能会怎样？人类赖以生存的石油、煤炭、天然气等化石能源以及众多矿藏都在哪里？这些都是研究者们期望通过"解码"去回答的问题。解读自然历史，旨在揭示自然规律，丰富人类知识宝库，并服务于国民经济建设。

饮水思源，早在 20 世纪早期，"南古人"* 便投身于化石的科学探索和研究中。这本图集展示了近百年来，一代又一代"南古人"对地层古生物学科的传承、贡献与开拓。

---

* "南古人"是现在或曾在南京古生物所（及其前身）工作的职工和学习的学生的简称，是他们自己对自己的称谓。

Postscript

Fossils, as depicted in this book, are truly natural treasures that have endured hundreds of million years of destructive processes, including weathering, chemical and biological decomposition, sedimentation, diagenesis, and orogenesis. They are rare "survivors" of the past biospheres, as the vast majority of all species, including their populations and individuals, of the deep time have vanished and must completely disintegrated into dust and ever smaller molecules for nature to recycle. They are here to reveal stories of the biosphere from the remote past, and through scientific examination, researchers can reconstruct scenes, events and evolution of the past biosphere. The fossils are also treasures of the palaeontological science as they are manifestation of interesting discoveries by Chinese palaeontologists.

The fossil specimens, that are now archived in the type repository and museum on the beautiful campus of Nanjing Institute of Geology and Palaeontology, Chinese Academy of Sciences (NIGPAS), represent a sample of collections made by several generations of NIGPAS and pre-NIGPAS researchers since the early 1920s, including the Institute's founding director Academician Professor Li Siguang (also known as J.S. Lee, early fusulinid expert), Professor Amadeus Grabau of Peking University, pioneers in studies of various fossil groups and academic sub-disciplines in China, and numerous other experts who devoted their whole life to fossil collection and research.

In retrospect, we greatly admire the untiring efforts made by earlier fossil collectors when the field work conditions were extremely hard and dangerous, even highly risky to the field explorers in early- and mid-twentieth century in China. The fossil displayed in this book are also windows of important academic achievements made by NIGPAS researchers during the past nearly 100 years. Through their arduous and groundbreaking works, not only has the palaeontological enterprise as a basic science branch taken shape, contributing to the world academic community, but also substantial input has been made to geological exploration of fossil fuels and mineral resources, as well as to general geology in China.

从南京古生物所的缔造者、中国第一代古生物学者的代表李四光先生，到各个古生物门类与研究领域的学术奠基人和中流砥柱，他们勇于开拓、敢于创新、励志攀登，取得了享誉世界的学术成果，堪称古生物学界的菁华。还有一批兢兢业业、默默无闻、锲而不舍，将自己的一生奉献给古生物研究的学者，他们不仅是许多重要科研成果的共同完成者，而且为地质、能源和矿产部门提供了强有力的科技支撑，亦可谓古生物学领域的精华。

这本图集采用的化石标本就是由这些科研人员不畏艰辛，于偏僻甚至是人迹罕至的野外采集到的。许多标本是采集者在早期极为艰苦的野外条件下，冒着生命危险获得的。这些化石标本大多都是模式标本，也就是这类化石生物第一次被人们发现并加以描述的标本。它们已被永久地保存在南京古生物所的标本馆中，成为固定的科学档案。

从科学规范上看，每一个模式标本所代表的化石名称，都是有充分的事实依据来支持的。若问标本的作用，简而言之："它们是科研成果的实证。"如果没有实证，科学结论往往就无据可依。

未来的科学研究仍然需要这些标本：需要运用新的研究技术和方法对标本进行再研究，以促进新的发现；国内外研究者发现的同类标本需要与已被研究的标本进行比较分析。

对于公众而言，这些科学档案中的许多化石都具有观赏价值，它们所蕴含的自然历史故事是引人入胜的，或能增进人们对大自然及其运行规律的了解。

采集这些化石标本影像的方法有很多种。有时只需要通过光源的配合，用普通照相机就能拍摄到精美的化石照片；有时则需要采用切片照相，才能看到利于研究的化石内部构造，比如珊瑚、苔藓虫、层孔虫等造礁生物的化石；有些化石特别微小，需要进行显微照相，比如单细胞的有孔虫、放射虫、硅藻、沟鞭藻、孢子花粉等化石；当普通的光学照相不足以达到研究目的时，可采用电子显微照相或激光共聚焦显微照相，以获得更高放大倍率和更大景深的电子或荧光影像；如果不想破坏化石整体而要观测内部结构的话，可以采用CT技术，即计算机断层成像技术，并采用虚拟切片的方式，把无数个不同焦距的平面影像叠加起来，达到"透视"的效果；最难以想象的是，把化石放到一个大型高能物理装置（同步辐射光源）中，从而获得分辨率很高的化石内部三维结构合成图像，只是这种大型科学装置造价高昂，在全世界都没有几个。

The fossil archives are important for a number of reasons. Firstly, the type specimens are material standards for the fossil species names that have been published by their discoverers; so when the same species is found elsewhere, the new material often need to be compared to the original type material to make sure that the new finding is correctly identified as the same species; or when later a dispute arises among researchers regarding the concept of the fossil species based on the original type specimens in our collections, the type material should be made available to the disputing parties for them to clarify. Secondly, when a new technique is available or a new idea occurs, the original type specimens are sometimes required to be re-examined, which often leads to new discoveries about the fossil species. Thirdly, many specimens in our collections are interesting to the general public and, thus, can make exhibitions for the public as a means of science education or general appreciation of the past organisms.

The images of fossils presented in this book were obtained via a number of techniques. Some are normal photographs of normal sized fossils taken by cameras via various lighting apparatus. Some are transmitted-light-photographs of thin sections cut through fossils like corals, bryozoans, stromatoporoids (often colonial reef animals), in order to observe the internal structures necessary for identifying the organisms. Some are micrographs taken under microscopes for microfossils such as the single-celled foraminifers, radiolarians, diatoms, dinoflagellates, and pollen and spores. For a greater depth of field in images, the scanning electron microscopy (SEM) is often used, in order to better show the microstructure of fossil organisms in high-magnification and higher clarity. More advance techniques, such as confocal fluorescence microscopy, micro-CT (for computerized tomography), and synchrotron tomography, are also often used for a virtual 3-D reconstruction of fossils without cutting and damaging the fossils. Most of the techniques and facilities are available in the laboratories of NIGPAS, but the synchrotron technique involves a huge high-energy physics facility which is currently only available elsewhere in very few locations in the world that are suitable for fossil imaging.

除了没有同步辐射光源，南京古生物所拥有上述其他几种采集化石影像的仪器设备。成像技术的不断发展和创新，给化石研究带来新的活力。这本图集展示的化石图像，就是通过以上设备采集并开展深入研究的，是众多化石图像的一个缩影。

出版这本图集的主要目的旨在介绍南京古生物所的化石藏品、悠久的古生物化石研究历史以及对国际学术界的贡献。我们希望借助这些影像信息的展示，引起同行和社会对化石的关注，并为公众特别是青少年，提供一些关于化石和古生物研究的简明信息和科学内涵。因此，我们没有强调化石门类和学术内容的完整性。

或许这本图集不能涵盖每一位读者所关注的内容，比如地球形成早期生命起源的相关信息以及人类起源的化石证据。其原因是，生命起源的问题还需要很多化学推论，较为抽象，难以用影像的形式表现出来；而人类起源的问题，由于南京古生物所的专业有限，相关化石标本更多地收藏在国内外其他科研机构或博物馆内。

我们特别感谢浙江大学出版社的编辑人员在编排、选图、文字和设计上的专业精神和认真友好的支持、帮助及合作。

编 者

2017 年夏，南京

This atlas is intended to introduce a sample of the fossil collections and the brief history of NIGPAS research development. Because the Institute is specialized in fossil invertebrates, plants and microfossils, we were not able to cover all aspects in palaeontology, especially palaeoanthropology and most of fossil vertebrates, for which excellent fossil specimens are deposited in many other museums and research institutes.

We thank the publishing team from Zhejiang University Press, who provided expert opinions and assistance in design, photographic style, and other aspects.

The Compiling Group
Nanjing, China
Summer, 2017

INTERNATIONAL CHRONO

www.stratigraphy.org

Eonothem / Eon	Erathem / Era	System / Period	Series / Epoch	Stage / Age	GSSP	numerical age (Ma)
Phanerozoic	Cenozoic	Quaternary	Holocene			present
						0.0117
			Pleistocene	Upper		0.126
				Middle		0.781
				Calabrian		1.80
				Gelasian		2.58
		Neogene	Pliocene	Piacenzian		3.600
				Zanclean		5.333
			Miocene	Messinian		7.246
				Tortonian		11.63
				Serravallian		13.82
				Langhian		15.97
				Burdigalian		20.44
				Aquitanian		23.03
		Paleogene	Oligocene	Chattian		27.82
				Rupelian		33.9
			Eocene	Priabonian		37.8
				Bartonian		41.2
				Lutetian		47.8
				Ypresian		56.0
			Paleocene	Thanetian		59.2
				Selandian		61.6
				Danian		66.0
	Mesozoic	Cretaceous	Upper	Maastrichtian		72.1 ±0.2
				Campanian		83.6 ±0.2
				Santonian		86.3 ±0.5
				Coniacian		89.8 ±0.3
				Turonian		93.9
				Cenomanian		100.5
			Lower	Albian		~ 113.0
				Aptian		~ 125.0
				Barremian		~ 129.4
				Hauterivian		~ 132.9
				Valanginian		~ 139.8
				Berriasian		~ 145.0

Eonothem / Eon	Erathem / Era	System / Period	Series / Epoch	Stage / Age	GSSP	numerical age (Ma)
						~ 145.0
Phanerozoic	Mesozoic	Jurassic	Upper	Tithonian		152.1 ±0.9
				Kimmeridgian		157.3 ±1.0
				Oxfordian		163.5 ±1.0
			Middle	Callovian		166.1 ±1.2
				Bathonian		168.3 ±1.3
				Bajocian		170.3 ±1.4
				Aalenian		174.1 ±1.0
			Lower	Toarcian		182.7 ±0.7
				Pliensbachian		190.8 ±1.0
				Sinemurian		199.3 ±0.3
				Hettangian		201.3 ±0.2
		Triassic	Upper	Rhaetian		~ 208.5
				Norian		~ 227
				Carnian		~ 237
			Middle	Ladinian		~ 242
				Anisian		247.2
			Lower	Olenekian		251.2
				Induan		251.902 ±0.024
	Paleozoic	Permian	Lopingian	Changhsingian		254.14 ±0.07
				Wuchiapingian		259.1 ±0.5
			Guadalupian	Capitanian		265.1 ±0.4
				Wordian		268.8 ±0.5
				Roadian		272.95 ±0.11
			Cisuralian	Kungurian		283.5 ±0.6
				Artinskian		290.1 ±0.26
				Sakmarian		295.0 ±0.18
				Asselian		298.9 ±0.15
		Carboniferous	Pennsylvanian Upper	Gzhelian		303.7 ±0.1
				Kasimovian		307.0 ±0.1
			Pennsylvanian Middle	Moscovian		315.2 ±0.2
			Pennsylvanian Lower	Bashkirian		323.2 ±0.4
			Mississippian Upper	Serpukhovian		330.9 ±0.2
			Mississippian Middle	Visean		346.7 ±0.4
			Mississippian Lower	Tournaisian		358.9 ±0.4

STRATIGRAPHIC CHART
Commission on Stratigraphy   v 2017/02

Eonothem / Eon	Erathem / Era	System / Period	Series / Epoch	Stage / Age	GSSP	numerical age (Ma)
Phanerozoic	Paleozoic	Devonian	Upper	Famennian	◄	358.9 ± 0.4
				Frasnian	◄	372.2 ±1.6
			Middle	Givetian	◄	382.7 ±1.6
				Eifelian	◄	387.7 ±0.8
						393.3 ±1.2
			Lower	Emsian		
				Pragian	◄	407.6 ±2.6
				Lochkovian	◄	410.8 ±2.8
						419.2 ±3.2
		Silurian	Pridoli		◄	423.0 ±2.3
			Ludlow	Ludfordian	◄	425.6 ±0.9
				Gorstian	◄	427.4 ±0.5
			Wenlock	Homerian	◄	430.5 ±0.7
				Sheinwoodian	◄	433.4 ±0.8
			Llandovery	Telychian	◄	438.5 ±1.1
				Aeronian	◄	440.8 ±1.2
				Rhuddanian	◄	443.8 ±1.5
		Ordovician	Upper	Hirnantian	◄	445.2 ±1.4
				Katian	◄	453.0 ±0.7
				Sandbian	◄	458.4 ±0.9
			Middle	Darriwilian	◄	467.3 ±1.1
				Dapingian	◄	470.0 ±1.4
			Lower	Floian	◄	477.7 ±1.4
				Tremadocian	◄	485.4 ±1.9
		Cambrian	Furongian	Stage 10		~ 489.5
				Jiangshanian	◄	~ 494
				Paibian	◄	~ 497
			Series 3	Guzhangian	◄	~ 500.5
				Drumian	◄	~ 504.5
				Stage 5		~ 509
			Series 2	Stage 4		~ 514
				Stage 3		~ 521
			Terreneuvian	Stage 2		~ 529
				Fortunian	◄	541.0 ±1.0

Eonothem / Eon	Erathem / Era	System / Period	GSSP GSSA	numerical age (Ma)
Precambrian	Proterozoic	Neo-proterozoic	Ediacaran	541.0 ±1.0
			Cryogenian	~ 635
			Tonian	~ 720
		Meso-proterozoic	Stenian	1000
			Ectasian	1200
			Calymmian	1400
		Paleo-proterozoic	Statherian	1600
			Orosirian	1800
			Rhyacian	2050
			Siderian	2300
	Archean	Neo-archean		2500
		Meso-archean		2800
		Paleo-archean		3200
		Eo-archean		3600
				4000
Hadean				~ 4600

Units of all ranks are in the process of being defined by Global Boundary Stratotype Section and Points (GSSP) for their lower boundaries, including those of the Archean and Proterozoic, long defined by Global Standard Stratigraphic Ages (GSSA). Charts and detailed information on ratified GSSPs are available at the website http://www.stratigraphy.org. The URL to this chart is found below.

Numerical ages are subject to revision and do not define units in the Phanerozoic and the Ediacaran; only GSSPs do. For boundaries in the Phanerozoic without ratified GSSPs or without constrained numerical ages, an approximate numerical age (~) is provided.

Numerical ages for all systems except Lower Pleistocene, Upper Paleogene, Cretaceous, Triassic, Permian and Precambrian are taken from 'A Geologic Time Scale 2012' by Gradstein et al. (2012); those for the Lower Pleistocene, Upper Paleogene, Cretaceous, Triassic, Permian and Precambrian were provided by the relevant ICS subcommissions.

Colouring follows the Commission for the Geological Map of the World (http://www.ccgm.org)

Chart drafted by K.M. Cohen, D.A.T. Harper, P.L. Gibbard
(c) International Commission on Stratigraphy, February 2017

To cite: Cohen, K.M., Finney, S.C., Gibbard, P.L. & Fan, J.-X. (2013; updated) The ICS International Chronostratigraphic Chart. Episodes 36: 199-204.

URL: http://www.stratigraphy.org/ICSchart/ChronostratChart2017-02.pdf

图书在版编目(CIP)数据

岁月菁华：化石档案与故事：汉英对照 / 中国科学院南京地质古生物研究所编；杨群主编. —杭州：浙江大学出版社，2017.12
ISBN 978-7-308-17629-3

Ⅰ.①岁… Ⅱ.①中… ②杨… Ⅲ.①古生物—化石—南京—画册 Ⅳ.①Q911.725.31—64

中国版本图书馆CIP数据核字(2017)第275096号

岁月菁华：化石档案与故事
中国科学院南京地质古生物研究所 编
杨　群　主编

策　　　划	徐有智　许佳颖
责任编辑	伍秀芳
封面设计	程　晨
装帧设计	程　晨　尤含悦
出版发行	浙江大学出版社
	（杭州市天目山路148号　邮政编码　310007）
	（网址：http://www.zjupress.com）
排　　版	杭州林智广告有限公司
印　　刷	浙江省邮电印刷股份有限公司
开　　本	889mm×1194mm　1/16
印　　张	15.25
字　　数	325千
版 印 次	2017年12月第1版　2017年12月第1次印刷
书　　号	ISBN 978-7-308-17629-3
定　　价	268.00 元

版权所有　翻印必究　　印装差错　负责调换
浙江大学出版社发行中心联系方式：(0571) 88925591；http://zjdxcbs.tmall.com